SpringerBriefs in Statistics

JSS Research Series in Statistics

The current research of statistics in Japan has expanded in several directions in line with recent trends in academic activities in the area of statistics and statistical sciences over the globe. The core of these research activities in statistics in Japan has been the Japan Statistical Society (JSS). This society, the oldest and largest academic organization for statistics in Japan, was founded in 1931 by a handful of pioneer statisticians and economists and now has a history of about 80 years. Many distinguished scholars have been members, including the influential statistician Hirotugu Akaike, who was a past president of JSS, and the notable mathematician Kiyosi Itô, who was an earlier member of the Institute of Statistical Mathematics (ISM), which has been a closely related organization since the establishment of ISM. The society has two academic journals: the Journal of the Japan Statistical Society (English Series) and the Journal of the Japan Statistical Society (Japanese Series). The membership of JSS consists of researchers, teachers, and professional statisticians in many different fields including mathematics, statistics, engineering, medical sciences, government statistics, economics, business, psychology, education, and many other natural, biological, and social sciences.

The JSS Series of Statistics aims to publish recent results of current research activities in the areas of statistics and statistical sciences in Japan that otherwise would not be available in English; they are complementary to the two JSS academic journals, both English and Japanese. Because the scope of a research paper in academic journals inevitably has become narrowly focused and condensed in recent years, this series is intended to fill the gap between academic research activities and the form of a single academic paper.

The series will be of great interest to a wide audience of researchers, teachers, professional statisticians, and graduate students in many countries who are interested in statistics and statistical sciences, in statistical theory, and in various areas of statistical applications.

More information about this series at http://www.springer.com/series/13497

Yan Liu · Fumiya Akashi
Masanobu Taniguchi

Empirical Likelihood and Quantile Methods for Time Series

Efficiency, Robustness, Optimality, and Prediction

 Springer

Yan Liu
Kyoto University/RIKEN AIP
Kyoto, Japan

Masanobu Taniguchi
Waseda University
Tokyo, Japan

Fumiya Akashi
Waseda University
Tokyo, Japan

ISSN 2191-544X   ISSN 2191-5458 (electronic)
SpringerBriefs in Statistics
ISSN 2364-0057   ISSN 2364-0065 (electronic)
JSS Research Series in Statistics
ISBN 978-981-10-0151-2 ISBN 978-981-10-0152-9 (eBook)
https://doi.org/10.1007/978-981-10-0152-9

Library of Congress Control Number: 2018963286

This Springer imprint is published by the registered company Springer Nature Singapore Pte Ltd.
The registered company address is: 152 Beach Road, #21-01/04 Gateway East, Singapore 189721, Singapore

To our families

Preface

There has been much demand for the statistical inference for dependent observations in many fields, such as economics, finance, biology, and engineering. A model that describes the probability structure of a series of dependent observations is called a stochastic process. The primary aim of this book is to discuss the nonparametric methods such as empirical likelihood method and quantile method for such stochastic processes. We deal with a wide variety of stochastic processes, for example, non-Gaussian linear processes, α-stable processes, and sinusoid models. We develop new estimation and hypothesis testing theory and discuss the efficiency, robustness, and optimality from the point of view of prediction problem for the stochastic processes. This book is designed for the researchers who specialize in or are interested in time series analysis.

Chapter 1 reviews the elements of stochastic processes. Especially, we introduce the Gaussian processes and α-stable processes for preparation. We also discuss the prediction problem, interpolation problem, and extrapolation problem for such processes in the frequency domain. Chapter 2 deals with the parameter estimation problem for stationary time series. We review several methods in minimum contrast estimation and formulate a new class of disparities for parameter estimation. The efficiency and robustness of the new disparities are discussed with numerical simulations. Chapters 3 and 4 focus on the nonparametric approach for time series analysis. The quantile methods in the frequency domain are discussed in Chap. 3. The scope of the methods is considered from second-order stationary processes to sinusoid models. Chapter 4 discusses the empirical likelihood methods for the pivotal quantity in the frequency domain. We investigate the asymptotic theory of the empirical likelihood ratio statistics for the linear processes with finite variance innovations and infinite variance innovations. Chapter 5 constructs another robust estimation/testing procedure, called self-weighting method for time series models. We generalize the empirical likelihood method to the self-weighted version and derive the pivotal limit distributions of the proposed statistics. Application of the GEL to the change point test of possibly infinite variance process is also discussed.

We have many people to thank. We thank Prof. Murad S. Taqqu for enlightening us to α-stable processes. We also thank Profs. Hans R. Künsch, Richard A. Davis,

Thomas Mikosch, and Claudia Klüppelberg for the suggestion on the statical inference for the infinite variance processes. We have also collaborated with Profs. Marc Hallin, Holger Dette, Roger Koenker, Ngai-Hang Chan, Anna C. Monti, Thomas Diccio, and Yoon-Jae Whang in the quantile and empirical likelihood methods for time series. Professors Yoshihide Kakizawa and Hiroaki Ogata provided us with high-level information for the empirical likelihood method for time series. We deeply thank all of these people. Finally, research by the first author was supported by JSPS Grant-in-Aid for Young Scientists (B) (17K12652), and the book was written while he was at Waseda University and belongs to Kyoto University and RIKEN AIP. Research by the second author was supported by JSPS Grant-in-Aid for Young Scientists (B) (16K16022) and was done at the Research Institute for Science & Engineering, Waseda University. Research by the third author was supported by JSPS Grant-in-Aid Kiban (A) (15H02081) and Kiban (S) (18H05290) and was done at the Research Institute for Science & Engineering, Waseda University.

Our thanks also go to the editor of JSS-Springer Series, Prof. Naoto Kunitomo, for his valuable comments to finalize the manuscripts into a book form. We also thank Mr. Yutaka Hirachi and Ms. Suganya Gnanamani for their great patience and cooperation in the final production of the book.

Kyoto, Japan   Yan Liu
Tokyo, Japan   Fumiya Akashi
Tokyo, Japan   Masanobu Taniguchi
November 2018

Contents

Chapter 1
Introduction

Abstract In this chapter, we describe some basic properties of stationary time series. Two fundamental approaches to time series analysis have been developed so far. One is the time-domain approach, and the other is the frequency-domain approach. In this book, we place the emphasis on the frequency-domain approach to analyzing stationary time series. Prediction problems, including interpolation and extrapolation problems, are discussed for stationary time series. In particular, we clarify the construction of a robust linear interpolator and extrapolator.

1.1 Stationary Time Series

A time series is supposed to be a realization of a family of random variables $\{X(t) : t \in T\}$. Such a family of random variables is called a *stochastic process*. The precise mathematical definition is as follows.

Definition 1.1 A stochastic process is a family of random variables $\{X(t) : t \in T\}$ defined on a probability space $(\Omega, \mathscr{F}, P)$.

In this book, we only consider the case that T is a countable set. Let $\mathscr{T}$ be the set of all vectors $\{t = (t_1, t_2, \ldots, t_n)^{\mathrm{T}} \in T^n : t_1 < t_2 < \cdots < t_n\}$. *The finite-dimensional joint distribution function* of the stochastic process $\{X(t)\}$ is defined as follows.

Definition 1.2 Consider the stochastic process $\{X(t) : t \in T\}$. For $t \in \mathscr{T}$, the function $F_t(\cdot)$, defined as

$$F_t(x) = P(X(t_1) < x_1, X(t_2) < x_2, \ldots, X(t_n) < x_n), \quad x = (x_1, x_2, \ldots, x_n)^{\mathrm{T}} \in \mathbb{R}^n,$$

is called the finite-dimensional joint distribution of $\{X(t)\}$. For any measurable function $g : \mathbb{R}^n \to \mathbb{R}^m$ of a random vector $X := \left(X(t_1), X(t_2), \ldots, X(t_n)\right)^{\mathrm{T}}$, the expectation of g is defined as

© The Author(s), under exclusive licence to Springer Nature Singapore Pte Ltd. 2018
Y. Liu et al., *Empirical Likelihood and Quantile Methods for Time Series*,
JSS Research Series in Statistics, https://doi.org/10.1007/978-981-10-0152-9_1

$$E[g(X)] = \int_{\mathbb{R}^n} g(x)\, dF_t(x).$$

The variance matrix of $g(X)$ is defined as

$$\text{Var}\,[g(X)] = E\Big(g(X) - E[g(X)]\Big)\Big(g(X) - E[g(X)]\Big)^{\mathsf{T}},$$

and the covariance of the random variables $X(t_1)$ and $X(t_2)$ is defined as

$$\text{Cov}\Big(X(t_1), X(t_2)\Big) = E\Big[\Big(X(t_1) - E[X(t_1)]\Big)\Big(X(t_2) - E[X(t_2)]\Big)\Big].$$

The following examples demonstrate some well-known stochastic processes.

Definition 1.3 (*Gaussian process*) A stochastic process $\{X(t)\}$ is called a *Gaussian process* if and only if all the finite-dimensional joint distributions of $\{X(t)\}$ are multivariate normal.

Definition 1.4 (*α-stable process*) A stochastic process $\{X(t)\}$ is called an *α-stable process* if and only if all the finite-dimensional joint distributions of $\{X(t)\}$ are multivariate α-stable distributions. (See Definition 1.5.)

For completeness, we introduce multivariate α-stable distributions. Let $X = (X_1, X_2, \ldots, X_d)^{\mathsf{T}} \in \mathbb{R}^d$ be a d-dimensional random vector and denote its characteristic function by $\phi(t)$, i.e.,

$$\phi(t) = E \exp(i t^{\mathsf{T}} X).$$

In addition, let S_d be a $(d-1)$-dimensional surface of a unit sphere in $\mathbb{R}^d$, i.e., $S_d = \{x \in \mathbb{R}^d : ||x|| = 1\}$.

Definition 1.5 Let $0 < \alpha < 2$. A random vector X is said to follow a *multivariate α-stable distribution* in $\mathbb{R}^d$ if and only if there exist a vector $\mu \in \mathbb{R}^d$ and a finite measure Γ on S_d such that the following hold:

(a) if $\alpha \neq 1$,

$$\phi(t) = \exp\left\{i t^{\mathsf{T}} \mu - \int_{S^d} |t^{\mathsf{T}} s|^{\alpha}\big(1 - i\,\text{sign}(t^{\mathsf{T}} s)\tan\big(\pi\alpha/2\big)\big)\Gamma(ds)\right\},$$

(b) if $\alpha = 1$,

$$\phi(t) = \exp\left\{i t^{\mathsf{T}} \mu - \int_{S^d} |t^{\mathsf{T}} s|\Big(1 + i\,\frac{2}{\pi}\text{sign}(t^{\mathsf{T}} s)\log|t^{\mathsf{T}} s|\Big)\Gamma(ds)\right\}.$$

For more on α-stable distributions, see Samoradnitsky and Taqqu (1994) and Nolan (2012).

Now, we introduce the concept of stationarity for a stochastic process. To be specific, let $T = \mathbb{Z}$. Such a stochastic process is referred to as a *time series*.

Definition 1.6 A time series $\{X(t) : t \in \mathbb{Z}\}$ is *strictly stationary* if the random vectors $(X(t_1), X(t_2), \ldots, X(t_k))^{\mathrm{T}}$ and $(X(t_1 + h), X(t_2 + h), \ldots, X(t_k + h))^{\mathrm{T}}$ have the same joint distribution for all positive integers k and all $t_1, t_2, \ldots, t_k, h \in \mathbb{Z}$.

As an alternative to strict stationarity, a useful concept is "second-order stationary" defined as follows.

Definition 1.7 A time series $\{X(t) : t \in \mathbb{Z}\}$ is *second-order stationary* if there exist a constant m and a sequence $\{\gamma(h) : h \in \mathbb{Z}\}$ such that

$$m = EX(t), \quad \text{for all } t \in \mathbb{Z},$$
$$\gamma(h) = \mathrm{Cov}(X(t), X(t + h)), \quad \text{for all } t, h \in \mathbb{Z},$$
$$\gamma(0) < \infty.$$

The function $\gamma(\cdot)$ is called *the autocovariance function* of the process $\{X(t)\}$.

Note that in Definition 1.7, a finite variance at any fixed time point is required for the time series. A strictly stationary process with finite second moments is second-order stationary. To facilitate a precise understanding of Definitions 1.6 and 1.7, we provide two examples.

Example 1.1 A second-order stationary Gaussian time series is strictly stationary.

Example 1.2 Even if an α-stable process is strictly stationary, if $\alpha \neq 2$, then the α-stable process does not satisfy the conditions in Definition 1.7, i.e., the strictly stationary α-stable process is not second-order stationary.

For the sake of brevity, we abbreviate "second-order stationary" as "stationary" hereafter. We introduce the nonnegative definiteness of a function as follows.

Definition 1.8 (*Nonnegative definiteness*) A real-valued function $\zeta : \mathbb{Z} \to \mathbb{R}$ is said to be *nonnegative definite* if and only if

$$\sum_{i,j=1}^{n} a_i \zeta(t_i - t_j) a_j \geq 0$$

for all positive integers n and all vectors $\boldsymbol{a} = (a_1, a_2, \ldots, a_n)^{\mathrm{T}} \in \mathbb{R}^n$ and $\boldsymbol{t} = (t_1, t_2, \ldots t_n)^{\mathrm{T}} \in \mathbb{Z}^n$.

The autocovariance function γ is nonnegative definite. To see this property, let $\boldsymbol{Y}_t = \left(X(t_1) - EX(t_1), X(t_2) - EX(t_2), \ldots, X(t_n) - EX(t_n) \right)^{\mathrm{T}}$ and $\Gamma_n = [\gamma(t_i - t_j)]_{i,j=1}^{n}$.

$$\sum_{i,j=1}^{n} a_i \zeta(t_i - t_j) a_j = \boldsymbol{a}^{\mathrm{T}} \Gamma_n \boldsymbol{a}$$

$$= \boldsymbol{a}^{\mathrm{T}} \left(E \, \boldsymbol{Y}_t \boldsymbol{Y}_t^{\mathrm{T}} \right) \boldsymbol{a}$$

$$= E \left(\boldsymbol{a}^{\mathrm{T}} \boldsymbol{Y}_t \right) \left(\boldsymbol{Y}_t^{\mathrm{T}} \boldsymbol{a} \right)$$

$$= \mathrm{Var} \left[\boldsymbol{a}^{\mathrm{T}} \boldsymbol{Y}_t \right]$$

$$\geq 0.$$

In addition, the spectral distribution function of a stationary time series $\{X(t)\}$ characterizes the behavior of the autocovariance function $\gamma(\cdot)$. Note that the autocovariance function γ can be regarded as a self-adjoint operator. Since every self-adjoint operator has an associated projection-valued measure, we obtain the following spectral theorem. This is well known as Herglotz's theorem.

Theorem 1.1 (Herglotz's theorem) *A complex-valued function γ defined on $\mathbb{Z}$ is nonnegative definite if and only if*

$$\gamma(h) = \int_{-\pi}^{\pi} \exp(ih\lambda) dF(\lambda) \quad \textit{for all } h \in \mathbb{Z},$$

where F is a right-continuous, nondecreasing, and bounded function on $\Lambda = [-\pi, \pi]$ and $F(-\pi) = 0$.

Let $\{X(t)\}$ be a zero-mean stationary process with a spectral distribution function F. An analytic comprehension of stationary processes is stated in the following spectral representation of $\{X(t)\}$.

Definition 1.9 (L^p-spaces) Let $(\mathscr{X}, \mathscr{B}, \mu)$ be a measure space and $0 < p \leq \infty$. $L^p(\mathscr{X}, \mathscr{B}, \mu)$ is defined as

$$L^p(\mathscr{X}, \mathscr{B}, \mu) = \left\{ f : \int_{\mathscr{X}} |f|^p \, d\mu < \infty \right\}, \quad \text{for } 0 < p < \infty,$$

and

$$L^\infty(\mathscr{X}, \mathscr{B}, \mu) = \left\{ f : \mu(|f| > K) = 0 \ \text{ for some } K \in (0, \infty) \right\}.$$

Here, the absolute value $|f|$ of a complex-valued function f is defined as $|f|^2 = f \bar{f}$, where $\bar{f}$ denotes the complex conjugate of f. In particular, when $\mathscr{X} = \Lambda$ is coupled with a measure F and its measurable sets, we abbreviate the notation as $L^p(F)$.

Definition 1.10 (*Closed linear span*) The closed span of any subset S of a space $\mathscr{X}$ is defined to be the smallest closed linear subspace $\mathscr{S}$ which is generated by the elements of S and contains all of the limit points under the norm of $\mathscr{X}$. Especially, we denote the closed linear span by sp, i.e., $\mathscr{S} = \mathrm{sp}\, S$.

Let $\mathcal{K} = \text{sp}\{X(t) : t \in \mathbb{Z}\}$ and $\mathcal{L} = \text{sp}\{\exp(it\cdot) : t \in \mathbb{Z}\}$ be subspaces in $L^2(\Omega, \mathcal{F}, P)$ and $L^2(F)$, respectively. There exists a unique linear isomorphism $\mathscr{I} : \mathcal{K} \to \mathcal{L}$, i.e., from the random variables in the time domain to the functions on $[-\pi, \pi]$ in the frequency domain. Let $Z(\lambda) = \mathscr{I}^{-1}(\mathbb{1}_{[-\pi, \lambda]}(\cdot))$ and define the linear isomorphism

$$\mathfrak{I}(f) = \int_{-\pi}^{\pi} f(\mu)\, Z(d\mu).$$

It can be shown that $\mathfrak{I} = \mathscr{I}^{-1}$ if we pay attention to each dense set of $L^2(\Omega, \mathcal{F}, P)$ and $L^2(F)$. The following spectral representation of $\{X(t)\}$ holds.

Theorem 1.2 *A zero-mean stationary process $\{X(t) : t \in \mathbb{Z}\}$ with a spectral distribution function F admits an expression in terms of a right-continuous orthogonal increment process $\{Z(\lambda), -\pi \leq \lambda \leq \pi\}$ such that*

(i) $X(t) = \int_{-\pi}^{\pi} \exp(-it\mu) Z(d\mu)$ *with probability one;*
(ii) $E|Z(\lambda) - Z(-\pi)|^2 = F(\lambda)$, *where $-\pi \leq \lambda \leq \pi$.*

The right-continuous orthogonal increment process $\{Z(\lambda), -\pi \leq \lambda \leq \pi\}$ has the following properties:

(i) $EZ(\lambda) = 0$;
(ii) $E|Z(\lambda)|^2 < \infty$;
(iii) $E\Big(Z(\lambda_4) - Z(\lambda_3)\Big)\overline{\Big(Z(\lambda_2) - Z(\lambda_1)\Big)} = 0$ if $(\lambda_1, \lambda_2] \cap (\lambda_3, \lambda_4] = \emptyset$;
(iv) $E|Z(\lambda + \delta) - Z(\lambda)|^2 \to 0$ as $\delta \to 0$.

This expression enables us to consider prediction problems for stationary processes in the frequency domain.

1.2 Prediction Problem

Let us consider prediction problem for a real-valued zero-mean stationary process $\{X(t) : t \in \mathbb{Z}\}$. For the process $\{X(t)\}$ with finite second moments, the optimality of predictors is often evaluated by the mean square error. To be precise and simple, let us first consider the 1-step ahead prediction problem. Without loss of generality, suppose we predict $X(0)$ by observations $\{X(t) : t \in S_1\}$, where $S_1 = \{x \in \mathbb{Z} : x \leq -1\}$. The prediction error under the mean square error is given by

$$\text{MSE}(\boldsymbol{a}) := E\left| X(0) - \sum_{j=1}^{\infty} a_j X(-j) \right|^2, \quad \boldsymbol{a} = (a_1, a_2, \ldots), \tag{1.1}$$

and the prediction problem is to solve the problem $\min_{\{a_i\}_{i=1}^{\infty}} \text{MSE}(\boldsymbol{a})$.

We interpret the 1-step ahead prediction problem in the frequency domain by the isomorphism $\mathscr{I}$. Applying the spectral representation of $\{X(t)\}$ in Theorem 1.2, we have

$$\min_{\{a_j\}_{j=1}^{\infty}} E\left|X(0) - \sum_{j=1}^{\infty} a_j X(-j)\right|^2 = \min_{\{a_j\}_{j=1}^{\infty}} E\left|\int_{-\pi}^{\pi} Z(d\lambda) - \sum_{j=1}^{\infty}\int_{-\pi}^{\pi} a_j \exp(ij\lambda)Z(d\lambda)\right|^2$$

$$= \min_{\{a_j\}_{j=1}^{\infty}} E\left|\int_{-\pi}^{\pi}\left(1 - \sum_{j=1}^{\infty} a_j \exp(ij\lambda)\right)Z(d\lambda)\right|^2$$

$$= \min_{\phi\in\mathscr{L}^2(S_1)} \int_{-\pi}^{\pi} |1 - \phi(\lambda)|^2 F(d\lambda), \qquad (1.2)$$

where $\phi(\lambda) = \sum_{j=1}^{\infty} a_j \exp(ij\lambda)$ and $\mathscr{L}^2(S_1)$ denotes the closed linear subspace of $L^2(F)$ generated by the set $\{e^{ij\lambda} : j \in S_1\}$. The third equality follows from the orthogonal increment property of the process $Z(\lambda)$.

To keep the argument simple, suppose there exists a spectral density f, the derivative of F, of the stationary process $\{X(t)\}$. Rewriting the right-hand side of (1.2), we obtain the 1-step ahead prediction problem

$$\min_{\phi\in\mathscr{L}^2(S_1)} \int_{-\pi}^{\pi} |1 - \phi(\lambda)|^2 f(\lambda)d\lambda. \qquad (1.3)$$

For the prediction problem, the following formula is well known. The formula is due to Szegö (1915) in the case that the spectral density f exists, with the extension to the general case (1.2) due to Kolmogorov in his papers Kolmogorov (1941b) and Kolmogorov (1941a). A polished extension is discussed in Koosis (1998). We substantially follow Brockwell and Davis (1991). The proof other than ours for (1.2) could be found in Helson and Lowdenslager (1958), whose approach is clearly exhibited in Hoffman (1962) and Helson (1964). See also Hannan (1970).

Theorem 1.3 (Kolmogorov's formula) *Suppose the spectral density f of the real-valued stationary process $\{X(t)\}$ is continuous on Λ and is bounded away from 0. The 1-step ahead prediction problem error of the process $\{X(t)\}$ is*

$$\sigma^2 := \min_{\phi\in\mathscr{L}^2(S_1)} \int_{-\pi}^{\pi} |1 - \phi(\lambda)|^2 f(\lambda)d\lambda = 2\pi \exp\left\{\frac{1}{2\pi}\int_{-\pi}^{\pi} \log f(\lambda)d\lambda\right\}. \quad (1.4)$$

Proof (Brockwell and Davis (1991)) Note that the Taylor expansion of $\log(1 - z)$ is

$$\log(1 - z) = -\sum_{j=1}^{\infty} \frac{z^j}{j}, \quad |z| < 1. \qquad (1.5)$$

We first suppose $f(\lambda)$ has the following expression, i.e.,

$$f(\lambda) = g(\lambda) := \frac{\sigma^2}{2\pi}\left|1 - g_1 e^{-i\lambda} - \cdots - g_p e^{-ip\lambda}\right|^{-2}, \qquad (1.6)$$

where $g(z) \equiv 1 - g_1 z - \cdots - g_p z^p \neq 0$ for $|z| \leq 1$. For any fixed p, we see that there exists a real constant M_g such that $\sum_{j=1}^{p} |g_j| \leq M_g$. At the same time, since the spectral density f in continuous on the compact set Λ, there exists a real constant M_f such that $f(\lambda) \leq M_f$ uniformly in λ. Thus, we have

$$\int_{-\pi}^{\pi} \left| \sum_{j=1}^{p} g_j e^{-ij\lambda} \right|^2 f(\lambda) d\lambda \leq \int_{-\pi}^{\pi} \sum_{j=1}^{p} \sum_{k=1}^{p} |g_j||g_k| f(\lambda) d\lambda$$
$$\leq 2\pi M_g^2 M_f.$$

Therefore, $\sum_{j=1}^{p} g_j e^{-ij\lambda} \in \mathscr{L}^2(S_1)$ and we obtain

$$\min_{\phi \in \mathscr{L}^2(S_1)} \int_{-\pi}^{\pi} |1 - \phi(\lambda)|^2 g(\lambda) d\lambda = \sigma^2.$$

A direct computation by (1.5) leads to

$$\int_{-\pi}^{\pi} \log g(\lambda) d\lambda = \int_{-\pi}^{\pi} \log \left(\frac{\sigma^2}{2\pi} \right) d\lambda - \int_{-\pi}^{\pi} \log \left| 1 - g_1 e^{-i\lambda} - \cdots - g_p e^{-ip\lambda} \right|^2 d\lambda$$
$$= 2\pi \log \left(\frac{\sigma^2}{2\pi} \right).$$

Thus, the right-hand side of (1.4) is

$$2\pi \exp \left\{ \frac{1}{2\pi} \int_{-\pi}^{\pi} \log g(\lambda) d\lambda \right\} = \sigma^2.$$

Now, we justify the expression (1.6) is sufficiently general. That is, for any positive ε and any symmetric continuous spectral density f bounded away from 0, there exists a spectral density g such that

$$|f(\lambda) - g(\lambda)| < \varepsilon, \quad \lambda \in [-\pi, \pi]. \tag{1.7}$$

Actually, from Corollary 4.4.2 in Brockwell and Davis (1991), the inequality (1.7) holds. Thus the set of $g(\lambda)$ is dense in the set of symmetric continuous spectral densities.

Returning back to (1.4), one can see that σ^2 is a linear functional of $f(\lambda)$. By the Cauchy–Schwarz inequality, it holds that for some fixed $M \in \mathbb{R}$,

$$\min_{\phi \in \mathscr{L}^2(S_1)} \int_{-\pi}^{\pi} |1 - \phi(\lambda)|^2 f(\lambda) d\lambda \leq \min_{\phi \in \mathscr{L}^2(S_1)} \left| \int_{-\pi}^{\pi} |1 - \phi(\lambda)|^4 d\lambda \right|^{1/2} \|f\|_{L^2} \leq M \|f\|_{L^2}.$$

The second inequality holds since $0 \in \mathscr{L}^2(S_1)$. Therefore, we obtain the second equality in (1.4) since it holds on a dense set in the set of symmetric continuous spectral densities, and the linear functional σ^2 is continuous. $\qquad\square$

An important result for the stationary process is the Wold decomposition. We leave the detail to other literature such as Hannan (1970) and Brockwell and Davis (1991). Let $\mathcal{M}_s$ be $\mathcal{M}_s = \text{sp}\{X(t) : -\infty < t \leq s\}$.

Theorem 1.4 (The Wold decomposition) *Suppose $\sigma^2 > 0$. A stationary process $\{X(t) : t \in \mathbb{Z}\}$ has the following unique expression:*

$$X(t) = \sum_{j=0}^{\infty} \psi_j Z(t-j) + V(t),$$

where

(i)   *$\psi_0 = 1$ and $\sum_{j=0}^{\infty} \psi_j^2 < \infty$;*
(ii) *$\{Z(t)\}$ is a sequence of uncorrelated random variables with variance σ^2;*
(iii) *for each $t \in \mathbb{Z}$, $Z(t) \in \mathcal{M}_t$;*
(iv) *for all $s, t \in \mathbb{Z}$, $E\,Z(t)V(s) = 0$;*
(v)   *for each $t \in \mathbb{Z}$, $V(t) \in \mathcal{M}_{-\infty}$;*
(vi) *$\{V(t)\}$ is deterministic.*

A linear time series is a useful model for time series analysis. The model is usually used in economics and finance (e.g., Box and Jenkins (1976), Hamilton (1994), Lütkepohl (2005) and Taniguchi and Kakizawa (2000)). Below are some concrete examples.

Example 1.3 The process $\{X(t) : t \in \mathbb{Z}\}$ is an ARMA(p, q) process if $\{X(t)\}$ is stationary and for any $t \in \mathbb{Z}$,

$$X(t) - b_1 X(t-1) - \cdots - b_p X(t-p) = Z(t) + a_1 Z(t-1) + \cdots + a_q Z(t-q),$$
$$\tag{1.8}$$

where $\{Z(t)\}$ is a sequence of zero-mean and uncorrelated random variables with variance σ^2. Especially, if $p = q = 0$, then the process is called *white noise*. For the ARMA process (1.8), the spectral density $f_X(\lambda)$ of $\{X(t)\}$ is

$$f_X(\lambda) = \frac{\sigma^2}{2\pi} \frac{|1 + \sum_{k=1}^{q} a_k e^{-ik\lambda}|^2}{|1 - \sum_{j=1}^{p} b_j e^{-ij\lambda}|^2}. \tag{1.9}$$

Several examples of the spectral function $F_X(\lambda)$ are given in Fig. 1.1.

Let us introduce the pth quantile λ_p of the spectral distribution function $F_X(\lambda)$. For simplicity, write $\Sigma_X := R_X(0)$. Note that the spectral distribution function $F_X(\lambda)$ takes values on $[0, \Sigma_X]$. The generalized inverse distribution function $F_X^{-1}(\psi)$ for $0 \leq \psi \leq \Sigma_X$ is defined as

$$F_X^{-1}(\psi) = \inf\{\lambda \in \Lambda ; F_X(\lambda) \geq \psi\}.$$

For $0 \leq p = \Sigma_X^{-1}\psi \leq 1$, we define the pth quantile λ_p as

$$\lambda_p := F_X^{-1}(p\Sigma_X) = \inf\{\lambda \in \Lambda \,;\, F_X(\lambda)\Sigma_X^{-1} \geq p\}. \qquad (1.10)$$

The statistical inference for the quantile λ_p is discussed in Chap. 3.

Example 1.4 The stationary process $\{X(t) : t \in \mathbb{Z}\}$ is a *linear process with finite variance innovations* if there exists a sequence $\{\psi_j\}$ such that $\sum_{j=0}^{\infty} |\psi_j| < \infty$ and

$$X(t) = \sum_{j=0}^{\infty} \psi_j Z(t - j), \qquad (1.11)$$

where $\{Z(t)\}$ is a sequence of zero-mean and uncorrelated random variables with variance σ^2. If $\{Z(t)\}$ is a sequence of i.i.d. random variables distributed as α-stable distribution and $\sum_{j=0}^{\infty} |\psi_j|^{\delta} < \infty$ for some $\delta \in (0, \alpha) \cap [0, 1]$, then the process $\{X(t)\}$ defined by (1.11) exists with probability 1 and is strictly stationary. In this case, we call the process $\{X(t)\}$ a *linear process with infinite variance innovations*.

Let us consider the existence of the generalized linear process $\{X(t)\}$ with infinite variance. First, suppose $1 < \alpha < 2$. In other words, $E|Z_t| < \infty$. Accordingly,

$$E|X(t)| \leq E\left(\sum_{j=0}^{\infty} |\psi_j Z(t - j)|\right) = \sum_{j=0}^{\infty} |\psi_j|\, E|Z(t - j)|$$

$$\leq \left\{\sum_{j=0}^{\infty} |\psi_j|^{\delta}\right\}^{1/\delta} E|Z(t)| < \infty.$$

Therefore, the process $\{X(t)\}$ is finite with probability 1.

Let us consider the other case of $0 < \alpha \leq 1$. Again, we can evaluate the absolute δth moment by the triangle inequality as follows:

$$E|X(t)|^{\delta} \leq E\left|\sum_{j=0}^{\infty} \psi_j Z(t - j)\right|^{\delta}$$

$$\leq E\sum_{j=0}^{\infty} |\psi_j|^{\delta} |Z(t - j)|^{\delta}$$

$$= \sum_{j=0}^{\infty} |\psi_j|^{\delta} E|Z(t)|^{\delta} < \infty,$$

since $\delta \in (0, \alpha)$. Therefore, the process $\{X(t)\}$ is finite with probability 1.

The statistical inference for the linear time series is described in the following chapters. Before that, we keep studying important properties of the stationary process. Especially, we consider the interpolation and extrapolation problem of stationary

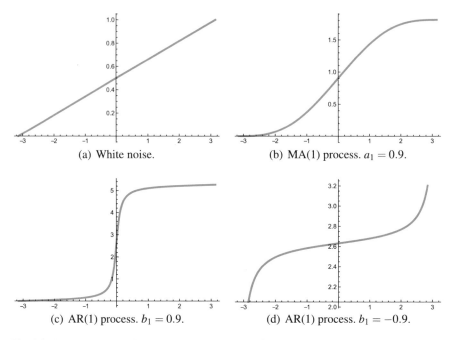

Fig. 1.1 Some examples of spectral distribution $F(\lambda)$ of stationary process $\{X(t)\}$

process, which stems from the prediction problem. The optimal interpolator and extrapolator are derived in both cases that the spectral density of the process is identified and not.

1.3   Interpolation and Extrapolation Problem

In this section, we consider interpolation and extrapolation problems of stationary processes. The interpolation and extrapolation problem is an extension of the concept of prediction for a stationary process. Some parts of this section are presented based on Liu et al. (2018).

The genuine interpolation problem for a stationary process is defined as follows. Suppose that all the values of the process $\{X(t) : t \in \mathbb{Z}\}$ are observed except for $X(0)$. The interpolation problem is to look for a linear combination $\hat{X}(0) = \sum_{j \neq 0} a_j X(j)$ that provides a good approximation of $X(0)$.

The extrapolation problem for a stationary process is more abstract. Suppose that some values before $X(0)$ of the process $\{X(t) : t \in \mathbb{Z}\}$ are observed. As in the previous section, define the set S_1 as $S_1 = \{x \in \mathbb{Z} : x \leq -1\}$. A precise formulation of the extrapolation problem is to assume that the values on the subset $U \subset S_1$ of the process $\{X(t)\}$ are observed. The extrapolation problem is then to look for a linear combination $\hat{X}(0) = \sum_{j \in U} a_j X(j)$ which provides a good approximation of $X(0)$. For instance, the 1-step ahead prediction problem is an extrapolation problem.

If we follow the consideration in Sect. 1.2 for the prediction problem, then we can see that both the interpolation and extrapolation problems can be formulated in the frequency domain. That is, the interpolation and extrapolation problems are to find $\hat{\phi}(\lambda)$ such that

$$\hat{\phi}(\lambda) = \arg \min_{\phi \in \mathscr{L}^2(S)} \int_{-\pi}^{\pi} |1 - \phi(\lambda)|^2 f(\lambda) d\lambda, \tag{1.12}$$

where S is an adequate set of observations from the stationary process $\{X(t)\}$.

The interpolation and extrapolation problems depend on the choice of the set S. This is described precisely in the following. Without loss of generality, we assume that $X(0)$ is always missing. Denote all the positive integers by $\mathbb{Z}_+$ and all the negative integers by $\mathbb{Z}_-$. We suppose that the observation set of the time series is given by $\{X(t) : t \in S\}$, where $S = S_- \cup S_+$, $S_- \subset \mathbb{Z}_-$ and $S_+ \subset \mathbb{Z}_+$. The interpolation problem corresponds to the case S satisfying $S_- \neq \emptyset$ and $S_+ \neq \emptyset$, while the extrapolation problem corresponds to the case that S ($\neq \emptyset$) satisfies $S_- = \emptyset$ or $S_+ = \emptyset$.

Several possible choices of the set S are given below.

Example 1.5 Some well-known choices of the set S are in the following:

- $S_1 = \{\ldots, -3, -2, -1\}$: the 1-step ahead prediction problem;
- $S_2 = \{\ldots, -3, -2, -1\} \cup \{1, 2, 3, \ldots\}$: the interpolation problem;
- $S_3 = \{\ldots, -3, -2, -1\} \setminus \{-k\}$: the incomplete past prediction problem;
- $S_4 = \{\ldots, -k-2, -k-1, -k\}$: the k-step ahead prediction problem.

Recall that the expression (1.12) stems from the spectral representation of $\{X(t)\}$ and the evaluation under the mean square error. Let us consider a more general problem, which is motivated by the work in Wald (1939) and Wald (1945). In the literature, it is suggested that the minimization problem

$$\min_{\theta} \int_{\mathbb{R}} L(\theta, x) P(dx) \tag{1.13}$$

is preferable, where $L(\theta, x)$ is a loss function of parameters and observations, and $P(x)$ is the distribution function of i.i.d. random variables. The reason why the minimization method (1.13) is preferable is that the distribution function P of i.i.d. random variables is uncertain in nature. Although the formulation (1.13) is considered for i.i.d. observations, it is also possible to formulate the procedure for a stationary process as follows.

For a stationary process $\{X(t)\}$, it is reasonable to regard the spectral distribution $F(\lambda)$ as "its distribution". Motivated by the equation (1.13), the extension of the mean square error to other functional norms is possible. From this point on, we discuss the following interpolation and extrapolation problem

$$\hat{\phi}(\lambda) = \arg \min_{\phi \in \mathscr{L}(S)} \int_{-\pi}^{\pi} |1 - \phi(\lambda)|^p f(\lambda) d\lambda, \quad p \geq 1, \tag{1.14}$$

where $\mathscr{L}(S)$ denotes the closed linear span of the set $\{e^{ij\lambda} : j \in S\}$ in $L^p(f)$. Under the formulation (1.14), we call $\hat{\phi}(\lambda)$ "*the optimal interpolator (or extrapolator)*".

To treat the minimization problem (1.14), we use a technique, known as the dual method, in the field of complex analysis. We first introduce the Hardy space as follows. Let $\mathscr{D}$ be the unit disk with center 0. The Hardy space $H^p(\mathscr{D})$ is a space consisting of all the holomorphic functions f on the open unit disk with

$$\left\{ \frac{1}{2\pi} \int_{-\pi}^{\pi} |f(re^{i\lambda})|^p d\lambda \right\}^{1/p} < \infty,$$

as $r \to 1^-$. Define the norm of f in $H^p(\mathscr{D})$ as

$$||f||_p = \lim_{r \to 1^-} \left\{ \frac{1}{2\pi} \int_{-\pi}^{\pi} |f(re^{i\lambda})|^p d\lambda \right\}^{1/p}.$$

From now on, we also write the function f using the expression $f(z) \equiv \sum_{k=0}^{\infty} f_k z^k$, $|z| < 1$, which is a function of a complex variable z, where $\{f_k\}$ denote the Fourier coefficients of the spectral density f.

We always suppose that q satisfies $1/p + 1/q = 1$ in the following to exploit the dual method. We discuss the derivation of the optimal extrapolator and interpolator in each case $S = S_4, S_1, S_2$, which are exemplified in Example 1.5.

Case $S = S_4$

Let us first consider the extrapolation problem of $S = S_4$, i.e., the minimization problem

$$\min_{\phi \in \mathscr{L}(S_4)} \int_{-\pi}^{\pi} |1 - \phi(\lambda)|^p f(\lambda) d\lambda. \tag{1.15}$$

Let $\mathscr{O}(z)$ be defined as

$$\mathscr{O}(z) = \exp\left(\frac{1}{2\pi p} \int_{-\pi}^{\pi} \frac{e^{i\lambda} + z}{e^{i\lambda} - z} \log f(e^{i\lambda}) d\lambda \right). \tag{1.16}$$

Thus, $\mathscr{O}^p$ is the outer function of f, and $|\mathscr{O}(e^{i\lambda})^p| = f(e^{i\lambda})$ almost everywhere (a.e.). (c.f. Hoffman (1962).) Suppose that the Fourier coefficients of $\mathscr{O}^{p/2}$ are given by $\{c_j\}$ and that the polynomial up to the $(k - 1)$th order is denoted by the polynomial operator p_k such that

$$p_k(\mathscr{O}^{p/2})(z) \equiv \sum_{j=0}^{k-1} c_j z^j.$$

Assume that the polynomial $p_k(\mathcal{O}^{p/2})$ has no zeros in $\mathcal{D}$. The expression of optimal extrapolator $\hat{\phi}$ is given in the following theorem.

Theorem 1.5 *Let* $S = S_4$. *Suppose that* f *is nonnegative and integrable*, $\log f \in L^1$. *The optimal extrapolator* $\hat{\phi}$ *of* (1.15) *is*

$$\hat{\phi}(\lambda) = 1 - \left(p_k^{2/p}(\mathcal{O}^{p/2})\right)(e^{-i\lambda}) \, \mathcal{O}^{-1}(e^{-i\lambda}) \ \ a.e.$$

The minimum error of (1.15) *is*

$$\min_{\phi \in \mathcal{L}(S_4)} \int_{-\pi}^{\pi} |1 - \phi(\lambda)|^p f(\lambda)d\lambda = \sum_{j=0}^{k-1} |c_j|^2.$$

Proof Let $\bar{S}_4$ be $\bar{S}_4 = \{k, k+1, k+2, \ldots\}$. It holds that $H^p(\mathcal{D}) = e^{-ik\lambda} \mathcal{L}(\bar{S}_4)$. We embed the problem (1.15) into the Hardy space $H^p(\mathcal{D})$ to find the optimal extrapolator $\hat{\phi}$ as follows. In fact,

$$\min_{\phi \in \mathcal{L}(S)} \left\{ \int_{-\pi}^{\pi} |1 - \phi(\lambda)|^p f(\lambda)d\lambda \right\}^{1/p}$$

$$= \min_{\phi \in \mathcal{L}(S_4)} \left\{ \int_{-\pi}^{\pi} |1 - \phi(\lambda)|^p f(\lambda)d\lambda \right\}^{1/p}$$

$$= \min_{\phi \in \mathcal{L}(S_4)} \left\{ \int_{-\pi}^{\pi} |e^{-ik\lambda} \mathcal{O}(e^{i\lambda}) - e^{-ik\lambda} \bar{\phi}(\lambda) \mathcal{O}(e^{i\lambda})|^p d\lambda \right\}^{1/p}$$

$$= \min_{g \in H^p(\mathcal{D})} \left\{ \int_{-\pi}^{\pi} |e^{-ik\lambda} \mathcal{O}(e^{i\lambda}) - g(\lambda)|^p d\lambda \right\}^{1/p} \quad \text{(say)}, \tag{1.17}$$

where we have the relation

$$g(\lambda) = e^{-ik\lambda} \bar{\phi}(\lambda) \mathcal{O}(e^{i\lambda}) \ \ a.e. \ \in H^p(\mathcal{D}). \tag{1.18}$$

Now, we focus on the function $e^{-ik\lambda} \mathcal{O}(e^{i\lambda})$ in (1.17). Let $\mathcal{P}$ be $\mathcal{P} = p_k(\mathcal{O}^{p/2})$. It holds that $z^{-k}\left(\mathcal{O}(z) - \mathcal{P}^{2/p}(z)\right) \in H^p(\mathcal{D})$. To see this, note that $\mathcal{O}^{p/2} - \mathcal{P}$ is analytic. Let us expand $\mathcal{O}^{p/2} - \mathcal{P}$ around the neighborhood of $z = 0$ as follows:

$$\mathcal{O}^{p/2}(z) - \mathcal{P}(z) = \sum_{n=0}^{\infty} a_n z^n,$$

where $a_0 = a_1 = \cdots = a_{k-1} = 0$. Then it also holds that

$$\mathcal{O}(z) - \mathcal{P}^{2/p}(z) = \sum_{n=0}^{\infty} b_n z^n,$$

where $b_0 = b_1 = \cdots = b_{k-1} = 0$, which can be shown by applying the chain rule as follows:

$$b_0 = \mathscr{O}(0) - \mathscr{P}^{2/p}(0)$$
$$= \mathscr{O}(0) - \left(\mathscr{O}(0)^{p/2} - a_0\right)^{2/p}$$
$$= 0,$$

$$b_1 = \mathscr{O}'(0) - (\mathscr{P}^{2/p})'(0)$$
$$= \frac{2}{p}\{\mathscr{O}^{p/2}(0)\}^{2/p-1}\left(\mathscr{O}^{p/2}\right)'(0) - \left\{\frac{2}{p}\mathscr{P}(0)^{2/p-1}\right\}\mathscr{P}'(0)$$
$$= \left\{\frac{2}{p}\mathscr{P}(0)^{2/p-1}\right\}\left\{\left(\mathscr{O}^{p/2}\right)'(0) - \mathscr{P}'(0)\right\}$$
$$= \left\{\frac{2}{p}\mathscr{P}(0)^{2/p-1}\right\}a_1$$
$$= 0,$$

and similarly, $b_2 = b_3 = \cdots = b_n = 0$ since $a_0 = a_1 = \cdots = a_{k-1} = 0$. Therefore, it holds that $z^{-k}(\mathscr{O} - \mathscr{P}^{2/p})(z) \in H^p(\mathscr{D})$.

Let us write $e^{-ik\lambda}\mathscr{O}(e^{i\lambda})$ in (1.17) as $e^{-ik\lambda}\mathscr{O}(e^{i\lambda}) = e^{-ik\lambda}\left(\mathscr{O}(e^{i\lambda}) - \mathscr{P}^{2/p}(e^{i\lambda})\right) + e^{-ik\lambda}\mathscr{P}^{2/p}(e^{i\lambda})$. Then the minimization problem (1.17) is equivalent to

$$\min_{g^* \in H^p(\mathscr{D})} \|e^{-ik\lambda}\mathscr{P}^{2/p} - g^*(\lambda)\|_p, \tag{1.19}$$

where

$$g^*(\lambda) = g(\lambda) - e^{-ik\lambda}\left(\mathscr{O}(e^{i\lambda}) - \mathscr{P}^{2/p}(e^{i\lambda})\right) \quad a.e. \tag{1.20}$$

From Theorem 8.1 in Duren (1970), it holds that

$$\min_{g^* \in H^p(\mathscr{D})} \|e^{-ik\lambda}\mathscr{P}^{2/p} - g^*\|_p = \max_{\substack{K(z) \in H^q(\mathscr{D}), \\ \|K(z)\|_q = 1}} \frac{1}{2\pi}\left|\int_{|z|=1} z^{-k}\mathscr{P}^{2/p}(z)K(z)dz\right|$$
$$= \max_{K(z) \in H^q(\mathscr{D})} \frac{\left|\int_{|z|=1} z^{-k}\mathscr{P}^{2/p}(z)K(z)dz\right|}{2\pi\|K\|_q}. \tag{1.21}$$

On the other hand, let $K^*(z)$ be

$$K^*(z) = \frac{z^k \mathscr{P}(z)\overline{\mathscr{P}(z)}}{\mathscr{P}^{2/p}(z)}.$$

For K^*, it holds that

$$||K^*||_q = \begin{cases} \frac{1}{2\pi}\left(\int_{|z|=1}\mathscr{P}(z)\overline{\mathscr{P}(z)}dz\right)^{1/q} & \text{when } p > 1; \\ \frac{1}{2\pi} & \text{when } p = 1. \end{cases}$$

We also have the expression

$$||e^{-ik\lambda}\mathscr{P}^{2/p}||_p = \frac{|\int_{|z|=1} z^{-k}\mathscr{P}^{2/p}(z)K^*(z)dz|}{2\pi||K^*||_q}. \tag{1.22}$$

Combining (1.21) and (1.22), we have

$$\begin{aligned}
||e^{-ik\lambda}\mathscr{P}^{2/p}||_p &\geq \min_{g^*\in H^p(\mathscr{D})} ||e^{-ik\lambda}\mathscr{P}^{2/p} - g^*||_p \\
&= \max_{K(z)\in H^q(\mathscr{D})} \frac{|\int_{|z|=1} z^{-k}\mathscr{P}^{2/p}(z)K(z)dz|}{2\pi||K||_q} \\
&\geq \frac{|\int_{|z|=1} z^{-k}\mathscr{P}^{2/p}(z)K^*(z)dz|}{2\pi||K^*||_q} \\
&= ||e^{-ik\lambda}\mathscr{P}^{2/p}||_p. \tag{1.23}
\end{aligned}$$

From (1.23), it can be seen that the function $g^* = 0$ is the minimizer of the extremal problem (1.19). From (1.18) and (1.20), we obtain that

$$\hat{\phi}(\lambda) = 1 - \left(p_k^{2/p}(\mathscr{O}^{p/2})\right)(e^{-i\lambda})\,\mathscr{O}^{-1}(e^{-i\lambda})\quad a.e., \tag{1.24}$$

which concludes Theorem 1.5.   □

Case $S = S_1$
Let $S = S_1 = \{\ldots, -3, -2, -1\}$. The case that $S = S_1$ is a special case of $S = S_4$ when $k = 1$. From (1.16), we obtain

$$\left(p_1^{2/p}(\mathscr{O}^{p/2})\right)(e^{i\lambda}) = c_0 = \exp\left(\frac{1}{2\pi p}\int_{-\pi}^{\pi} \log f(e^{i\lambda})d\lambda\right).$$

From (1.24), we have

$$|1 - \hat{\phi}(\lambda)|^p = \frac{1}{f(\lambda)}\exp\left(\frac{1}{2\pi}\int_{-\pi}^{\pi} \log f(\lambda)d\lambda\right).$$

Hence, it holds that

$$\min_{\phi\in\mathscr{L}(S)}\int_{-\pi}^{\pi}|1 - \phi(\lambda)|^p f(\lambda)d\lambda = 2\pi\exp\left\{\frac{1}{2\pi}\int_{-\pi}^{\pi} \log f(\lambda)d\lambda\right\}. \tag{1.25}$$

Equation (1.25) shows nothing but Kolmogorov's formula in Theorem 1.3.

Case $S = S_2$

Next, let us consider the case that $S = S_2 = \{\ldots, -3, -2, -1\} \cup \{1, 2, 3, \ldots\}$, i.e., the interpolation problem

$$\min_{\phi \in \mathscr{L}(S_2)} \int_{-\pi}^{\pi} |1 - \phi(\lambda)|^p f(\lambda) d\lambda. \tag{1.26}$$

The expression of the optimal interpolator $\hat{\phi}$ is given in the following theorem.

Theorem 1.6 *Let* $S = S_2$. *The optimal interpolator* $\hat{\phi}$ *of* (1.26) *is*

$$\hat{\phi}(\lambda) = 1 - \frac{(2\pi) f(\lambda)^{-q/p}}{\int_{-\pi}^{\pi} f(\lambda)^{-q/p} d\lambda} \quad a.e. \tag{1.27}$$

(i) If $p > 1$ and $f^{-q/p} \in L^s$ for some $s > 1$, then the minimum error of (1.26) *is*

$$\min_{\phi \in \mathscr{L}(S_2)} \int_{-\pi}^{\pi} |1 - \phi(\lambda)|^p f(\lambda) d\lambda = (2\pi)^p \left(\int_{-\pi}^{\pi} f(\lambda)^{-q/p} d\lambda \right)^{-p/q}.$$

(ii) If $p = 1$, then the minimum error of (1.26) *is*

$$\min_{\phi \in \mathscr{L}(S_2)} \int_{-\pi}^{\pi} |1 - \phi(\lambda)|^p f(\lambda) d\lambda = \frac{2\pi}{\|f^{-1}\|_{L^\infty}}.$$

Proof Let us rewrite the minimum as follows.

$$\min_{\phi \in \mathscr{L}(S_2)} \int_{-\pi}^{\pi} |1 - \phi(\lambda)|^p f(\lambda) d\lambda = \min_{\phi \in \mathscr{L}(S_2)} \frac{\int_{-\pi}^{\pi} |1 - \phi(\lambda)|^p f(\lambda) d\lambda}{(2\pi)^{-p} |\int_{-\pi}^{\pi} (1 - \phi(\lambda)) d\lambda|^p}. \tag{1.28}$$

The right-hand side of (1.28) is scale invariant if we regard it as a function of $1 - \phi$. We embed ϕ into the space $\mathscr{L}(S_2 \cup \{0\})$ and change the problem into a maximization problem as

$$\min_{\phi \in \mathscr{L}(S_2)} \frac{\int_{-\pi}^{\pi} |1 - \phi(\lambda)|^p f(\lambda) d\lambda}{(2\pi)^{-p} |\int_{-\pi}^{\pi} (1 - \phi(\lambda)) d\lambda|^p}$$

$$= \min_{g \in \mathscr{L}(S_2 \cup \{0\})} \frac{\int_{-\pi}^{\pi} |g(\lambda)|^p f(\lambda) d\lambda}{(2\pi)^{-p} |\int_{-\pi}^{\pi} g(\lambda) d\lambda|^p}$$

$$= \left(\max_{g \in \mathscr{L}(S_2 \cup \{0\})} \frac{(2\pi)^{-p} |\int_{-\pi}^{\pi} g(\lambda) d\lambda|^p}{\int_{-\pi}^{\pi} |g(\lambda)|^p f(\lambda) d\lambda} \right)^{-1}$$

$$= \left(\max_{g \in \mathscr{L}(S_2 \cup \{0\})} \frac{(2\pi)^{-p} |\int_{-\pi}^{\pi} g(\lambda) f(\lambda)^{1/p} f(\lambda)^{-1/p} d\lambda|^p}{\int_{-\pi}^{\pi} |g(\lambda) f(\lambda)^{1/p}|^p d\lambda} \right)^{-1}. \tag{1.29}$$

From (1.28) and (1.29), it holds that

$$\min_{\phi \in \mathscr{L}(S_2)} \left(\int_{-\pi}^{\pi} |1 - \phi(\lambda)|^p f(\lambda) d\lambda \right)^{1/p}$$

$$= 2\pi \left(\max_{g \in \mathscr{L}(S_2 \cup \{0\})} \frac{\int_{-\pi}^{\pi} g(\lambda) f(\lambda)^{1/p} f(\lambda)^{-1/p} d\lambda}{\|gf^{1/p}\|_{L^p}} \right)^{-1}. \quad (1.30)$$

If $f^{-q/p} \in L^s$ for some $s > 1$, then $f^{-1/p} \in L^q$. Thus, from Theorem 7.1 in Duren (1970), we have

$$\max_{g \in \mathscr{L}(S_2 \cup \{0\})} \frac{\int_{-\pi}^{\pi} g(\lambda) f(\lambda)^{1/p} f(\lambda)^{-1/p} d\lambda}{\|gf^{1/p}\|_{L^p}}$$

$$= \min_{\substack{g^*: \int_{-\pi}^{\pi} gf^{1/p}g^* d\lambda = 0 \\ \forall g \in \mathscr{L}(S_2 \cup \{0\}), \int_{-\pi}^{\pi} g d\lambda \neq 0}} \|f^{-1/p} - g^*\|_{L^q}. \quad (1.31)$$

Let us consider the constraint condition for the minimization problem of the right-hand side in (1.31). Observing that $\int_{-\pi}^{\pi} gf^{1/p}g^* d\lambda = 0$ holds for any $g \in \mathscr{L}(S_2 \cup \{0\})$ and $\int_{-\pi}^{\pi} g d\lambda \neq 0$, we come to the conclusion that $g^* = 0$ a.e. Thus, from (1.30) and (1.31), we have, if $p > 1$, then

$$\min_{\phi \in \mathscr{L}(S_2)} \int_{-\pi}^{\pi} |1 - \phi(\lambda)|^p f(\lambda) d\lambda = (2\pi)^p \left(\int_{-\pi}^{\pi} f(\lambda)^{-q/p} d\lambda \right)^{-p/q};$$

if $p = 1$, then

$$\min_{\phi \in \mathscr{L}(S_2)} \int_{-\pi}^{\pi} |1 - \phi(\lambda)|^p f(\lambda) d\lambda = \frac{2\pi}{\|f^{-1}\|_{L^\infty}}.$$

In addition, the optimal interpolator $\hat{\phi}$ is

$$\hat{\phi}(\lambda) = 1 - \frac{(2\pi) f(\lambda)^{-q/p}}{\int_{-\pi}^{\pi} f(\lambda)^{-q/p} d\lambda} \quad a.e.$$

The existence of $\hat{\phi}$ is guaranteed by Carleson's theorem in Hunt (1968). In fact, if $f^{-q/p} \in L^s$ for some $s > 1$, then it holds that

$$\lim_{N \to \infty} \sum_{n=-N}^{N} \widehat{f^{-q/p}}(n) \cdot e^{in\lambda} = f(\lambda)^{-q/p} \quad a.e.,$$

where $\widehat{f^{-q/p}}$ is the Fourier coefficient of the function $f^{-q/p}$. This concludes Theorem 1.6. □

Example 1.6 Suppose that $p = 2$ in the case of $S = S_2$. This is the classical interpolation problem for stationary time series, which is evaluated under the mean square error. The optimal interpolator $\hat{\phi}$ is

$$\hat{\phi}(\lambda) = 1 - f(\lambda)^{-1} \left(\frac{1}{2\pi} \int_{-\pi}^{\pi} f(\lambda)^{-1} d\lambda \right)^{-1}.$$

The minimum interpolation error is

$$\min_{\phi \in \mathscr{L}(S_2)} \int_{-\pi}^{\pi} |1 - \phi(\lambda)|^2 f(\lambda) d\lambda = 4\pi \left(\int_{-\pi}^{\pi} f(\lambda)^{-1} d\lambda \right)^{-1}.$$

The extrapolation and interpolation problems are important subjects from a practical viewpoint. However, the assumption that the true spectral density f is known is not natural in practice. The practitioner is required to estimate the true spectral density f through some statistical methods and make use of extrapolation and interpolation strategies after the estimation. The statistical estimation for the spectral density f will be discussed in the next chapter.

1.4   Robust Interpolation and Extrapolation

In this section, we discuss robust interpolation and extrapolation problems. We have seen that the optimal interpolator and extrapolator require the knowledge of the true spectral density. Here, it is discussed here how we can deal with cases in which we do not have the full knowledge of the spectral density. For this problem, it is usual to assume that the spectral density f is contained in a contaminated class.

Let $\mathscr{D}$ denote the class of all spectral densities supported by the interval Λ, where the integral on Λ is 1. Suppose we have knowledge of a spectral density function h which belongs to a class $\mathscr{F}$ of ε-contaminated spectral densities, that is,

$$\mathscr{F} = \{h \in \mathscr{D} : h = (1 - \varepsilon)f + \varepsilon g, \ g \in \mathscr{D}\}, \quad 0 < \varepsilon < 1, \tag{1.32}$$

where f is a fixed spectral density and g denotes a function with uncertainty which ranges over the class $\mathscr{D}$.

We define *the minimax linear interpolator (or extrapolator)* by ϕ^*, which minimizes the maximal interpolation error (or extrapolation error) with respect to h, i.e.,

$$\max_{h \in \mathscr{F}} \int_{-\pi}^{\pi} |1 - \phi^*(\lambda)|^p h(\lambda) d\lambda = \min_{\phi \in \mathscr{L}(S)} \max_{h \in \mathscr{F}} \int_{-\pi}^{\pi} |1 - \phi(\lambda)|^p h(\lambda) d\lambda, \tag{1.33}$$

where $\mathscr{L}(S)$ denotes the intersection $\cap_{h \in \mathscr{F}} \mathscr{L}^h(S)$. Here, $\mathscr{L}^h(S)$ is the closed linear span of the set $\{e^{ij\lambda} : j \in S\}$ in $L^p(h)$ for any spectral density $h \in \mathscr{F}$. We adopt the minimax principle as the criterion of the optimality.

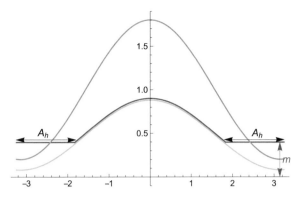

Fig. 1.2   The yellow line denotes the spectral density f, the green line represents $(1 - \varepsilon)f$, and the blue line shows the least favorable spectral density h^*. The set A_h and $m := \text{ess.} \inf_{\lambda \in \Lambda} h(\lambda)$ are indicated by the black arrows and red arrow. As shown for interpolation and extrapolation problems, $\mathscr{G} \neq \emptyset$ (Condition 1) if the known spectral density f is bounded away from 0. The least favorable spectral density h^* exists if there exists a spectral density $g^* \in \mathscr{G}$ (Condition 2)

Let ess. sup f (or ess. inf f) denote the essential supremum (or infimum) of the function f, and let μ be the Lebesgue measure on Λ. For any fixed spectral density h, let $\hat{\phi}_h$ be the optimal interpolator (or extrapolator) with respect to the spectral density h, which has already been discussed in Sect. 1.3, and let the set $A_h(\subset \Lambda)$ be

$$A_h = \{\lambda \in \Lambda : |1 - \hat{\phi}_h(\lambda)|^p = \text{ess.} \sup_{\lambda \in \Lambda} |1 - \hat{\phi}_h(\lambda)|^p\}. \tag{1.34}$$

In addition, define a class $\mathscr{G}$ of the uncertainty spectral density g as

$$\mathscr{G} = \{g \in \mathscr{D} : \mu(A_h) > 0 \text{ for } h = (1 - \varepsilon)f + \varepsilon g\}. \tag{1.35}$$

In general, the minimax interpolation and extrapolation problem (1.33) can be solved under the following two conditions:

Condition 1. For the spectral density f and uncertainty class $\mathscr{D}$, $\mathscr{G} \neq \emptyset$.

Condition 2. For the spectral density f, there exists a spectral density $g^* \in \mathscr{G} \subset \mathscr{D}$ with $h^* = (1 - \varepsilon)f + \varepsilon g^*$, whose support is the set A_{h^*}.

In fact, the spectral density h^* is *the least favorable spectral density* in class $\mathscr{F}$ of (1.32). An example of the least favorable spectral density is shown in Fig. 1.2.

Under Conditions 1 and 2, the optimal interpolator (or extrapolator) under uncertainty exists, and it is described by the following theorem.

Theorem 1.7 *If Conditions 1 and 2 hold, then we have*

$$\min_{\phi \in \mathscr{L}(S)} \max_{h \in \mathscr{F}} \int_{-\pi}^{\pi} |1 - \phi(\lambda)|^p h(\lambda) d\lambda = \int_{-\pi}^{\pi} |1 - \hat{\phi}_{h^*}(\lambda)|^p h^*(\lambda) d\lambda. \tag{1.36}$$

The optimal interpolator (or extrapolator) ϕ^* under uncertainty satisfies $\phi^* = \hat{\phi}_{h^*}$.

Proof Let us denote the interpolation (or extrapolation) error for the interpolator (or extrapolator) ϕ with respect to the spectral density h by

$$e(\phi, h) = \int_{-\pi}^{\pi} |1 - \phi(\lambda)|^p h(\lambda) d\lambda.$$

First, let us consider the maximization problem $\max_{h \in \mathscr{F}} e(\phi, h)$. From the property of the class $\mathscr{F}$ in (1.32), this is equivalent to considering the maximization problem

$$(1 - \varepsilon) e(\phi, f) + \varepsilon \max_{g \in \mathscr{D}} e(\phi, g), \tag{1.37}$$

since the spectral density f is fixed. From Hölder's inequality, for any $\phi \in \mathscr{L}(S)$ it holds that

$$e(\phi, g) \leq |||1 - \phi|^p||_{L^\infty} ||g||_{L^1}. \tag{1.38}$$

Under Condition 1, $e(\phi, h)$ can attain its upper bound if $g \in \mathscr{G}$; otherwise, there is no such $h \in \mathscr{F}$. Note that the equation in (1.38) holds if the support of g is the set $A_h = \{\lambda \in \Lambda : |1 - \hat{\phi}_h(\lambda)|^p = \text{ess.sup}_\lambda |1 - \hat{\phi}_h|^p\}$. Now, from Condition 2, we have

$$\max_{g \in \mathscr{D}} e(\hat{\phi}_{h^*}, g) = e(\hat{\phi}_{h^*}, g^*). \tag{1.39}$$

Thus, from (1.37) and (1.39), it holds that

$$\max_{h \in \mathscr{F}} e(\hat{\phi}_{h^*}, h) = (1 - \varepsilon) e(\hat{\phi}_{h^*}, f) + \varepsilon \max_{g \in \mathscr{D}} e(\hat{\phi}_{h^*}, g) = e(\hat{\phi}_{h^*}, h^*). \tag{1.40}$$

In addition, from the definition of the interpolator (or extrapolator), it holds that

$$e(\hat{\phi}_{h^*}, h^*) = \min_{\phi \in \mathscr{L}(S)} e(\phi, h^*). \tag{1.41}$$

Combining (1.40) and (1.41), we have

$$\max_{h \in \mathscr{F}} e(\hat{\phi}_{h^*}, h) = e(\hat{\phi}_{h^*}, h^*) = \min_{\phi \in \mathscr{L}(S)} e(\phi, h^*).$$

Furthermore, for arbitrary $\phi \in \mathscr{L}(S)$, it holds that

$$\max_{h \in \mathscr{F}} e(\phi, h) \geq e(\phi, h^*) \geq \min_{\phi \in \mathscr{L}(S)} e(\phi, h^*) = \max_{h \in \mathscr{F}} e(\hat{\phi}_{h^*}, h). \tag{1.42}$$

Consequently, if we take the minimum with respect to $\phi \in \mathscr{L}(S)$ on both sides in (1.42), then

$$\min_{\phi \in \mathscr{L}(S)} \max_{h \in \mathscr{F}} \int_{-\pi}^{\pi} |1 - \phi(\lambda)|^P h(\lambda) d\lambda = \int_{-\pi}^{\pi} |1 - \hat{\phi}_{h^*}(\lambda)|^P h^*(\lambda) d\lambda.$$

holds, since $\hat{\phi}_{h^*} \in \mathscr{L}(S)$, which concludes the equation (1.36) in Theorem 1.7. □

Example 1.7 Let us consider the minimax extrapolation problem for the case of $S = S_1$ in Example 1.5. In other words, we consider the minimax 1-step ahead prediction problem. For an arbitrary spectral density $h \in \mathscr{F}$, it follows from Theorem 1.5 that

$$|1 - \hat{\phi}_h(\lambda)|^P = \frac{1}{h(\lambda)} \exp\left(\frac{1}{2\pi} \int_{-\pi}^{\pi} \log h(\lambda) d\lambda\right) \quad a.e. \tag{1.43}$$

We now show that there exists a unique spectral density $h^* \in \mathscr{F}$ such that h^* satisfies Conditions 1 and 2. From the definition of the set A_h in (1.34) and the right-hand side of (1.43), an equivalent form of the set A_h is

$$A_h = \left\{\lambda \in \Lambda : |1 - \hat{\phi}_h(\lambda)|^P = \text{ess.} \sup_{\lambda \in \Lambda} \frac{1}{h(\lambda)} \exp\left(\frac{1}{2\pi} \int_{-\pi}^{\pi} \log h(\lambda) d\lambda\right)\right\}$$
$$= \{\lambda \in \Lambda : h(\lambda) = \text{ess.} \inf_{\lambda \in \Lambda} h(\lambda)\}.$$

For an arbitrary spectral density $h \in \mathscr{F}$, let m, E_m and F_m be defined as

$$m = \text{ess.} \inf_{\lambda \in \Lambda} h(\lambda), \quad E_m = A_h, \quad \text{and} \quad F_m = \Lambda - A_h.$$

Suppose that there exists a least favorable spectral density h^* such that $\mu(E_m) > 0$. We show that there exists no contradiction.

Since $\mu(E_m) > 0$, let g^* be distributed on the set E_m. Thus, g^* should be

$$g^*(\lambda) = \begin{cases} 0 & \text{for, } \lambda \in F_m, \\ \frac{m - (1 - \varepsilon) f(\lambda)}{\varepsilon} & \text{for } \lambda \in E_m, \end{cases}$$

since $\text{ess.} \inf_{\lambda \in \Lambda} h(\lambda) = m$. From the fact that $g^* \in \mathscr{D}$, that is, the integral of g^* is 1, m should satisfy the following equation:

$$\int_{E_m} \frac{m - (1 - \varepsilon) f(\lambda)}{\varepsilon} d\lambda = 1. \tag{1.44}$$

Also, from the facts that g^* is nonnegative, $\mu(E_m) > 0$ and (1.43), the concrete characterization of the decomposition of the interval Λ, by E_m and F_m, should be

$$E_m = \{\lambda \in \Lambda : m \geq (1 - \varepsilon) f(\lambda)\}; \tag{1.45}$$
$$F_m = \{\lambda \in \Lambda : m < (1 - \varepsilon) f(\lambda)\}.$$

Noting that the left-hand side of equation (1.44), $(1/\varepsilon)\int_{E_m} m - (1-\varepsilon)f(\lambda)d\lambda$, is an increasing and continuous function with respect to m in the domain $[0, \infty)$, the solution to the determination of m satisfying (1.44) and (1.45) exists uniquely. Thus, the spectral density $h^*(\lambda)$ satisfying Conditions 1 and 2, i.e.,

$$h^*(\lambda) = \begin{cases} (1-\varepsilon)f(\lambda) & \text{for, } \lambda \in F_m, \\ m & \text{for } \lambda \in E_m, \end{cases} \qquad (1.46)$$

exists uniquely. Clearly, $\mu(E_m) > 0$ when $\varepsilon > 0$. From Theorem 1.7, the predictor $\hat{\phi}_{h^*}$ is the minimax 1-step ahead linear predictor. The solution in Hosoya (1978) is a special case of our result for $p = 2$.

Example 1.8 Next, let us consider the minimax interpolation problem for the case that $S = S_2$ in Example 1.5. From Theorem 1.6, it holds that

$$|1 - \hat{\phi}_h(\lambda)|^p = \frac{(2\pi)^p h^{-q}(\lambda)}{(\int_{-\pi}^{\pi} h^{-q/p}(\lambda)d\lambda)^p} \quad a.e. \qquad (1.47)$$

Again, the set A_h is equivalent to the following definition:

$$A_h = \{\lambda \in \Lambda : h(\lambda) = \text{ess. } \inf_{\lambda \in \Lambda} h(\lambda)\}.$$

From Example 1.7, there exists a unique spectral density $h^* \in \mathscr{F}$ such that h^* satisfies Conditions 1 and 2 and h^* is defined by (1.46). Therefore, from Theorem 1.7, the minimax linear interpolator ϕ^* is $\hat{\phi}_{h^*}$. We omit the details since the proof is similar to the case that $S = S_1$ in Example 1.7. In particular, the result in Taniguchi (1981b) is a special case of ours for $p = 2$.

The above discussion is specialized to the case that the spectral distribution H with uncertainty is absolutely continuous with respect to the Lebesgue measure. When this is not the case, we must modify the optimal interpolator (or extrapolator) ϕ^*. A detailed discussion can be found in Liu et al. (2018). It is shown that for the modified interpolator (or extrapolator) $\tilde{\phi}$, the following inequality:

$$\max_H \int_{-\pi}^{\pi} |1 - \tilde{\phi}(\lambda)|^p h(\lambda)d\lambda < \int_{-\pi}^{\pi} |1 - \hat{\phi}_{h^*}(\lambda)|^p h^*(\lambda)d\lambda + \delta \qquad (1.48)$$

holds for any $\delta > 0$. The equation (1.48) shows that there exists an approximate interpolator (or extrapolator) $\tilde{\phi}$ in the case that the spectral distribution H is not absolutely continuous with respect to the Lebesgue measure such that the interpolation error (or extrapolation error) of $\tilde{\phi}$ is arbitrarily close to that of ϕ^* when the spectral density h exists.

Finally, we show some numerical comparisons for the robust interpolation problem. Note that when the spectral density h is fixed, it follows from (1.47) that

$$|1 - \hat{\phi}_h(\lambda)|^p h(\lambda) = \frac{(2\pi)^p h^{1-q}(\lambda)}{\left(\int_{-\pi}^{\pi} h^{-q/p}(\lambda) d\lambda\right)^p},$$

for any $p \geq 1$. Suppose that the uncertain spectral density g is contained in the following subfamily $\mathscr{D}_s$, i.e.,

$$\mathscr{D}_s = \{g \in \mathscr{D} : g(\lambda) = (2\pi)^{-k} k(\lambda + \pi)^{k-1}, \ k \in \mathbb{R}^+\}, \tag{1.49}$$

and hence, the spectral density h is

$$h \in \mathscr{F}_s = \{h \in \mathscr{D} : (1 - \varepsilon)f + \varepsilon g, \ g \in \mathscr{D}_s\} \subset \mathscr{F},$$

where $\mathbb{R}^+$ denotes the positive real numbers. Note that the spectral density for a real-valued stationary process is symmetric about the origin. We do not assume that this is the case in our numerical studies. If one is interested in a real-valued stationary process, then one can reduce the function along the x-axis direction to $[0, \pi]$, and reverse the function on $[0, \pi]$ to $[-\pi, 0]$ with respect to the y-axis.

Now, we change the parameter ε, p and the fixed spectral density f to see the least favorable case in numerical studies in the following settings:

(i) fix $p = 2$ and $f(\lambda) = 1/(2\pi)$ for $\lambda \in \Lambda$. $\varepsilon = 0.1, 0.2, \cdots, 0.9$;
(ii) fix $f(\lambda) = 1/(2\pi)$ for $\lambda \in \Lambda$ and $\varepsilon = 0.2$. $p = 1.2, 1.4, \ldots, 3.0$;
(iii) fix $\varepsilon = 0.2$ and $p = 2$. Let the spectral density f take the form f_l as follows, i.e.,

$$f_l(\lambda) = (2\pi)^{-l} l(\lambda + \pi)^{l-1}, \tag{1.50}$$

where we change the parameter l as $l = 0.80, 0.85, \ldots, 1.20$;
(iv) fix $\varepsilon = 0.2$ and $p = 1.2$. Let the spectral density f take the same form (1.50) as $l = 0.80, 0.85, \ldots, 1.20$.

For the setting (i), let us first derive the least favorable spectral density h^*. To specify m in (1.46), we focus on the decomposition of Λ as in (1.45). Here, $E_m = \Lambda$ since f is constant and $\mu(E_m) \neq 0$. From (1.44), it holds that

$$(2\pi)m - (1 - \varepsilon) = \varepsilon,$$

and hence $m = 1/(2\pi)$. Thus, the least favorable spectral density h^* is

$$h^*(\lambda) = f(\lambda) = \frac{1}{2\pi}. \tag{1.51}$$

From (1.51), we observe that $g^*(\lambda) = 1/(2\pi) \in \mathscr{D}_s$. It is the least favorable case when $k = 1$. The numerical results are illustrated in Fig. 1.3.

For the setting (ii), we fix $\varepsilon = 0.2$. The least favorable spectral density h^* is the same as that in (1.51) from Example 1.8. Similarly, it is the least favorable case when $k = 1$. The numerical results are displayed in Fig. 1.4. We do not include the case of

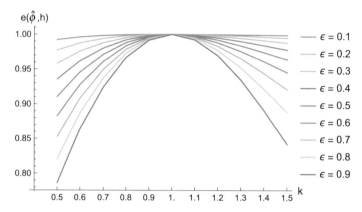

Fig. 1.3 For setting (i), the optimal interpolation error takes its maximum when $k = 1$. Accordingly, the least favorable spectral density h^* is what we have shown in (1.51)

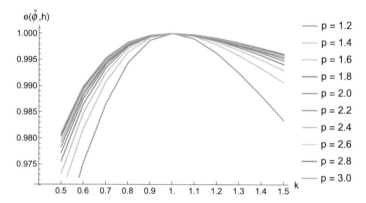

Fig. 1.4 For setting (ii), the optimal interpolation error takes its maximum when $k = 1$. Accordingly, the least favorable spectral density h^* is also what we have shown in (1.51)

$p = 1$ in our numerical results since Condition 1 is not satisfied for the case $p = 1$ as we consider the subfamily (1.49).

In the setting (iii), we fix $\varepsilon = 0.2$ and $p = 2$ and change the parameter l of the spectral density f_l. We explain how to determine the parameter m of the least favorable spectral density h^* for each l except for $l = 1$ in the following example.

Example 1.9 When the spectral density f takes the form (1.50), it is obvious that the spectral density f is continuous on $(-\pi, \pi)$. As what we have considered in Example 1.7, except for the case that $l = 1$, the spectral density f_l and m have an intersection point and let it be λ_m. Hence, if $0 < l < 1$, then m and λ_m satisfy

$$\begin{cases} (1 - \varepsilon) \int_{-\pi}^{\lambda_m} f_l(\lambda)\, d\lambda + m(\pi - \lambda_m) = 1, \\ (1 - \varepsilon) f_l(\lambda_m) = m, \end{cases} \tag{1.52}$$

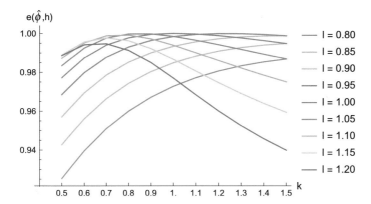

Fig. 1.5 For setting (iii), the least favorable spectral density h^* is not contained in the subfamily $\mathscr{F}_s$, except for the case that $l = 1$. This setting shows different aspects from settings (i) and (ii) in that the constant uncertain spectral density $(k = 1)$ is not the least favorable one in the subfamily $\mathscr{D}_s$

if $l > 1$, then m and λ_m satisfy

$$\begin{cases} (1 - \varepsilon) \int_{\lambda_m}^{\pi} f_l(\lambda)\, d\lambda + m(\pi + \lambda_m) = 1, \\ (1 - \varepsilon) f_l(\lambda_m) = m. \end{cases} \tag{1.53}$$

Note that Eqs. (1.52) and (1.53) are not linear and they require some numerical computations.

Now, let us focus on the subfamily (1.49) of uncertain spectral densities g. Note that it is impossible to find the least favorable spectral density h^* of (1.46) from the subfamily $\mathscr{D}_s$, except for the case that $l = 1$. That is to say, if $l \neq 1$, then $\bar{g} = 1/(2\pi)$ does not make the spectral density h the least favorable. Numerically, this can be found from Fig. 1.5.

It can also be shown by finding a spectral density $\tilde{g} \in \mathscr{D}_s$ which makes $\tilde{h}$ dominate $\bar{h}$ in the sense that

$$e(\hat{\phi}_{\tilde{h}}, \tilde{h}) > e(\hat{\phi}_{\tilde{h}}, \bar{h}), \tag{1.54}$$

where $\tilde{h}$ and $\bar{h}$ are

$$\tilde{h} = (1 - \varepsilon) f_l + \varepsilon \tilde{g},$$
$$\bar{h} = (1 - \varepsilon) f_l + \varepsilon \bar{g}.$$

Let us consider the next example.

Example 1.10 In the case that $p = 2$, it follows from Theorem 1.6 that the optimal interpolation error $e(\hat{\phi}_h, h)$ for the stationary process $\{X(t)\}$ with spectral density h is

$$e(\hat{\phi}_h, h) = (2\pi)^2 \Big(\int_{-\pi}^{\pi} h(\lambda)^{-1} \, d\lambda \Big)^{-1}.$$

To make the computation simple, let $l = 2$, that is,

$$f(\lambda) = f_2(\lambda) = (2\pi)^{-2} 2(\lambda + \pi).$$

Note that the function $h = 1/(2\pi)$ is not contained in the class $\mathscr{F}$ if $\varepsilon \neq 0.5$.

Suppose $k = 1/2$ and then $\tilde{g} = (2\pi)^{-1/2}(\lambda + \pi)^{-1/2}/2$. As above, we compare the optimal interpolation errors of the spectral density functions $\tilde{h}$ and $\bar{h}$, i.e.,

$$\tilde{h} = (1 - \varepsilon)(2\pi)^{-2} 2(\lambda + \pi) + \varepsilon(2\pi)^{-1/2}(\lambda + \pi)^{-1/2}/2,$$
$$\bar{h} = (1 - \varepsilon)(2\pi)^{-2} 2(\lambda + \pi) + \varepsilon/(2\pi),$$

and show that (1.54) holds. In fact, for the spectral density $\tilde{h}$, it holds that

$$e(\hat{\phi}_{\tilde{h}}, \tilde{h}) = (2\pi)^2 \Big(\int_{-\pi}^{\pi} \tilde{h}(\lambda)^{-1} \, d\lambda \Big)^{-1}$$
$$= (2\pi)^2 \Big(\int_{-\pi}^{\pi} 2 \big/ \big((1 - \varepsilon)(2\pi)^{-2} 4(\lambda + \pi) + \varepsilon(2\pi)^{-1/2}(\lambda + \pi)^{-1/2}\big) d\lambda \Big)^{-1}$$
$$= 3(1 - \varepsilon) \Big/ \log\Big(1 + \frac{4(1 - \varepsilon)}{\varepsilon}\Big). \tag{1.55}$$

On the other hand, for the spectral density $\bar{h}$, we have

$$e(\hat{\phi}_{\bar{h}}, \bar{h}) = (2\pi)^2 \Big(\int_{-\pi}^{\pi} \bar{h}(\lambda)^{-1} \, d\lambda \Big)^{-1}$$
$$= (2\pi)^2 \Big(\int_{-\pi}^{\pi} 1 \big/ \big((1 - \varepsilon)(2\pi)^{-2} 2(\lambda + \pi) + \varepsilon/(2\pi)\big) d\lambda \Big)^{-1}$$
$$= 2(1 - \varepsilon) \Big/ \log\Big(1 + \frac{2(1 - \varepsilon)}{\varepsilon}\Big). \tag{1.56}$$

Easily, it can be found that two functions (1.55) and (1.56) of ε intersect at $\varepsilon = 1 - 1/\sqrt{5}$, and numerically, $\varepsilon \approx 0.553$. When $\varepsilon < 1 - 1/\sqrt{5}$, (1.54) always holds. In other words, when the fixed spectral density f is different from the constant function, then the case that g is constant ($k = 1$) is not always the least favorite.

Finally, we focus on the setting (iv), i.e., $p \neq 2$. In particular, we fix $p = 1.2$. In this case, it follows from Theorem 1.6 again that the optimal interpolation error $e(\hat{\phi}_h, h)$ is

$$e(\hat{\phi}_h, h) = (2\pi)^{6/5} \Big(\int_{-\pi}^{\pi} h(\lambda)^{-5} d\lambda \Big)^{-1/5}. \tag{1.57}$$

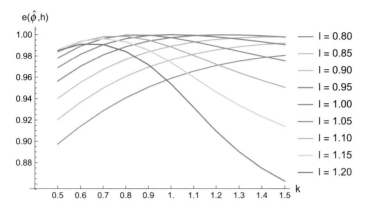

Fig. 1.6 The numerical results of setting (iv) are similar to those of the setting (iii). The least favorable spectral density h^* is not contained in the subfamily $\mathscr{F}_s$, except for the case that $l = 1$. In addition, the constant uncertain spectral density $(k = 1)$ is not the least favorable one in the subfamily $\mathscr{D}_s$

Again, it is possible to proceed as in Example 1.10 to discuss that the case that g is constant $(k = 1)$ is not always the least favorite. However, the analytical computation of (1.57) is extremely complex so we leave it to the reader. The numerical results for several cases are shown in Fig. 1.6.

In summary, when we have some knowledge of the true spectral density with some uncertainty, we can apply the methodology described in this section to finding a robust predictor or a robust interpolator. To obtain the knowledge of the true spectral density from the observations of a stationary process, we need to make a statistical inference for our data. The parametric method for spectral inference is discussed in Chap. 2, while the nonparametric method is discussed in Chap. 3.

Chapter 2
Parameter Estimation Based on Prediction

Abstract In this chapter, we discuss the parameter estimation problem for stationary time series. The method of maximum likelihood is usually not tractable since the joint distribution of the time series does not have analytical expression even if the model is identified. Instead, an alternative method is to estimate the parameters of the time series model by minimizing contrast function. We introduce a new class of contrast functions for parameter estimation from the prediction problem, which is quite different from other existing literature. Under both settings for a stationary process with finite and infinite variance innovations, we investigate asymptotic properties of such minimum contrast estimators. It is shown that the contrast function corresponding to the prediction error is asymptotically the most efficient in the class. In our simulation, we compare the relative efficiency and robustness against randomly missing observations for different contrast functions among the class.

2.1 Introduction

We have reviewed the basic properties of stationary time series so far. The numerical data with dependence relation can be generated from the time series models we have discussed. In reality, we often do not have knowledge of the mechanism, especially the mathematical model, which generates what we have observed. For this purpose, we have to specify the mathematical models from observations of the stochastic process. One fundamental method is to fit parametric linear models to time series data. By the isomorphism in Sect. 1.1, the specification of the spectral density of stationary time series is equivalent to that of the model. Thus, we discuss parameter estimation of spectral density function in this chapter.

The method of maximum likelihood has many optimal properties such as full asymptotic efficiency among regular estimators for parameter estimation from the point of view of mathematical statistics. The method is applicable to the Gaussian time series since all the finite-dimensional joint distributions of stochastic process are multivariate normal. From this desirable property, (Whittle 1952a, 1953, 1954) systematically investigated the parameter estimation for the Gaussian process. Especially, he suggested an approximate Gaussian likelihood for parameter estimation

© The Author(s), under exclusive licence to Springer Nature Singapore Pte Ltd. 2018
Y. Liu et al., *Empirical Likelihood and Quantile Methods for Time Series*,
JSS Research Series in Statistics, https://doi.org/10.1007/978-981-10-0152-9_2

of spectral density, which is well known as the Whittle likelihood afterward. The method he proposed is to approximate the inverse of the covariance matrix of the Gaussian process by the spectral density, and then estimate the parameters in the spectral density.

Suppose $\{X(1), X(2), \ldots, X(n)\}$ is an observation stretch from stationary process $\{X(t) : t \in \mathbb{Z}\}$. From Theorem 1.1, the process has a spectral distribution function F. Let us suppose there exists a spectral density f for the process.

Let $I_{n,X}(\omega)$ denote the periodogram of the observations, that is,

$$I_{n,X}(\omega) = \frac{1}{2\pi n} \left| \sum_{t=1}^{n} X(t) e^{it\omega} \right|^2.$$

To specify the spectral density f of the process, a natural idea makes us suppose the observation stretch $\{X(1), X(2), \ldots, X(n)\}$ comes from an ARMA(p, q) process, which has the spectral density

$$f_\theta(\omega) = \frac{\sigma^2}{2\pi} \frac{|1 + \sum_{k=1}^{q} a_k e^{-ik\omega}|^2}{|1 - \sum_{j=1}^{p} b_j e^{-ij\omega}|^2},$$

where $\theta = (a_1, a_2, \ldots, a_q, b_1, b_2, \ldots, b_p)$ is the parameter to specify (cf. (1.8)). The Whittle likelihood is defined as

$$\int_{-\pi}^{\pi} I_{n,X}(\omega) / f_\theta(\omega) d\omega. \tag{2.1}$$

The Whittle estimator $\hat{\theta}_n$ minimizes the Whittle likelihood (2.1). Although the Whittle likelihood is originally an approximation of the Gaussian log likelihood, the estimation procedure does not require one to assume the process is Gaussian. This is one reason why the Whittle likelihood is popular for parameter estimation of stationary time series model.

The method of minimizing contrast function, on the other hand, is motivated by robust estimation, when there is uncertainty about the probability model. Some typical examples of contrast functions are distance or disparity measures between observations and models. In time series analysis, the contrast function $D(f_\theta, \hat{g}_n)$ is introduced for parameter estimation, where f_θ is a parametric spectral density and $\hat{g}_n$ is a nonparametric estimator for spectral density constructed by observations. For example, if we take $\hat{g}_n = I_{n,X}$, then Eq. (2.1) can be understood by

$$D(f_\theta, I_{n,X}) = \int_{-\pi}^{\pi} I_{n,X}(\omega) / f_\theta(\omega) d\omega.$$

The criterion must enjoy desirable properties. One is that the true parameter θ_0 should be the minimizer of the contrast function $D(f_\theta, f)$, where f is the true spectral density function of the stationary time series. Some disparity measures as follows have already been introduced in the literature.

2.1.1 Location Disparity

One well-known disparity is location disparity, which is introduced in Taniguchi (1981a). The location disparity is given by

$$D_l(f_\theta, \hat{g}_n) = \int_{-\pi}^{\pi} \Phi(f_\theta(\omega))^2 - 2\Phi(f_\theta(\omega))\Phi(\hat{g}_n(\omega))d\omega,$$

where $\hat{g}_n$ is a nonparametric estimator of the spectral density and Φ is an appropriate bijective function. Several examples of the bijective function $\Phi(\cdot)$ are

(i) $\Phi(x) = \log x,$
(ii) $\Phi(x) = 1.$

The first example (i) of function $\Phi(\cdot)$ is Bloomfield's exponential model, which is proposed in Bloomfield (1973). The advantage of the model is that the true parameter θ_0, which minimize the criterion $D_l(f_\theta, f)$, can be easily expressed in an explicit form.

2.1.2 Scale Disparity

Another contrast function, scale disparity, is studied in Taniguchi (1987). The discussion on the higher order asymptotics of the disparity can be found in Taniguchi et al. (2003). The criterion is given as

$$D_s(f_\theta, \hat{g}_n) = \int_{-\pi}^{\pi} K(f_\theta(\omega)/\hat{g}_n(\omega))d\omega,$$

where K is sufficiently smooth with its minimum at 1. Without loss of generality, the scale function K can also be modified by some function $\tilde{K}$ that

$$\tilde{K}\left(\frac{f_\theta(\omega)}{\hat{g}_n(\omega)} - 1\right) := K\left(\frac{f_\theta(\omega)}{\hat{g}_n(\omega)}\right),$$

where the minimum of $\tilde{K}$ is at 0. Some examples of K are

  (i) $K(x) = \log x + 1/x,$
 (ii) $K(x) = -\log x + x,$
(iii) $K(x) = (\log x)^2,$
(iv) $K(x) = (x^\alpha - 1)^2, \quad \alpha \neq 0,$
 (v) $K(x) = x \log x - x,$
(vi) $K(x) = \log\{(1 - \alpha) + \alpha x\} - \alpha \log x, \quad \alpha \in (0, 1).$

The scale disparity, in fact, has much broader use than the location disparity. Examples (i) and (iii) have equivalent expressions to what we considered in the location case, respectively. Example (vi) is investigated in Albrecht (1984) as an α-entropy criterion for the Gaussian process. This contrast function is robust with respect to the peak, which is discussed by Zhang and Taniguchi (1995).

2.1.3 A New Disparity Based on Prediction

We introduce a new class of contrast functions for parameter estimation from the prediction problem. A general form of the new contrast functions is defined as follows:

$$D_p(f_\theta, f) = \int_{-\pi}^{\pi} \left(\int_{-\pi}^{\pi} f_\theta(\omega)^{\alpha+1} d\omega \right)^{\beta} f_\theta(\omega)^{\alpha} f(\omega) d\omega, \quad \alpha \neq 0, \ \beta \in \mathbb{R}. \quad (2.2)$$

The disparity is introduced for parameter estimation of time series models in Liu (2017a). Comparing with the disparity D_l and D_s, we can see that D_p is not included in either location disparity or scale disparity since D_l is a measure of the difference between the true density and the model transformed by a bijective function, and D_s is a measure of the difference between the ratio of the true density to the model and 1. On the other hand, as for the disparity (2.2), the model f_θ and the true density f have the different power in its own form.

To present the result in general, we need some preparation. Let θ be in a parameter space $\Theta \subset \mathbb{R}^d$. Suppose the parametric spectral density f_θ is bounded away from 0. The disparity (2.2) is motivated by the following two examples concerning the prediction error and the interpolation error.

Example 2.1 (Prediction error) Removing the constant term when $\alpha = -1$, we obtain the following disparity:

$$D_p(f_\theta, f) = \int_{-\pi}^{\pi} f(\omega)/f_\theta(\omega) d\omega. \quad (2.3)$$

If we substitute the periodogram $I_{n,X}$ for the true spectral density f in (2.3), then it is called the Whittle likelihood. This disparity relates to the prediction error when we use the model f_θ to make a prediction. Actually, by the same argument in Sect. 1.2, from (1.1) and (1.2), we can see that the prediction error in general is expressed as

$$\int_{-\pi}^{\pi} |1 - \phi(\omega)|^2 f(\omega) d\omega,$$

where $\phi(\omega) = \sum_{j=1}^{\infty} a_j e^{-ij\omega}$. If we take the inverse $f_\theta(\omega)^{-1} = |1 - \phi(\omega)|^2$ with $\theta = (a_1, a_2, \ldots)$, then $f_\theta(\omega)$ is just a spectral density for the AR model without the constant multiple. Consequently, we can interpret the Whittle likelihood as a disparity standing for the prediction error when we fit the model f_θ.

Example 2.2 (Interpolation error) If we plug $\alpha = -2$ and $\beta = -2$ into the disparity (2.2), we obtain

$$D_p(f_\theta, f) = \left(\int_{-\pi}^{\pi} \frac{1}{f_\theta(\omega)} d\omega\right)^{-2} \left(\int_{-\pi}^{\pi} \frac{f(\omega)}{f_\theta(\omega)^2} d\omega\right). \tag{2.4}$$

This disparity (2.4) was investigated in Suto et al. (2016). The disparity relates to the interpolation error when we use the model f_θ to interpolate.

Recall that the interpolation error is expressed by

$$\int_{-\pi}^{\pi} |1 - \phi(\omega)|^2 f(\omega) d\omega, \tag{2.5}$$

where $\phi(\omega) = \sum_{j \neq 0} a_j e^{ij\omega}$. To minimize (2.5), $1 - \phi$ must be orthogonal to $\mathscr{L}^2(S_2)$ under $L^2(f)$-norm. In other words, we have

$$\int_{-\pi}^{\pi} (1 - \phi(\omega)) f(\omega) e^{ik\omega} d\omega = 0, \quad \forall k \in S_2.$$

Thus, there exists a constant $c \in \mathbb{C}$ such that

$$(1 - \phi(\omega)) f(\omega) = c. \tag{2.6}$$

However, noting that $\phi(\omega) \in \mathscr{L}^2(S_2)$, we have

$$\int_{-\pi}^{\pi} |1 - \phi(\omega)|^2 f(\omega) d\omega = \int_{-\pi}^{\pi} (1 - \phi(\omega)) f(\omega) \overline{(1 - \phi(\omega))} d\omega = 2\pi c,$$

which shows that $c \in \mathbb{R}$ since the interpolation error must be real.

In addition, since the spectral density $f(\omega)$ is a real-valued function, it holds that

$$\begin{aligned} 2\pi c &= \int_{-\pi}^{\pi} |1 - \phi(\omega)|^2 f(\omega) d\omega \\ &= \int_{-\pi}^{\pi} \left((1 - \phi(\omega)) f(\omega)\right) f(\omega)^{-1} \overline{f(\omega)(1 - \phi(\omega))} d\omega \\ &= c^2 \int_{-\pi}^{\pi} f(\omega)^{-1} d\omega, \end{aligned}$$

which shows that

$$c = \left(\frac{1}{2\pi} \int_{-\pi}^{\pi} f(\omega)^{-1} d\omega \right)^{-1} \tag{2.7}$$

except for the trivial case. Now, if we insert (2.7) into (2.6), we obtain the optimal interpolator as

$$\phi(\omega) = 1 - f(\omega)^{-1} \left(\frac{1}{2\pi} \int_{-\pi}^{\pi} f(\omega)^{-1} d\omega \right)^{-1}. \tag{2.8}$$

Note that we have already obtained this result by a more general formula (1.23) in Sect. 1.3, if we let $p = q = 2$ in formula (1.27). (See also Example 1.6.)

Now, we replace the true density $f(\omega)$ by $f_\theta(\omega)$ in (2.8) to see the relation between the interpolation error and the disparity (2.4). That is,

$$\int_{-\pi}^{\pi} |1 - \phi(\omega)|^2 f(\omega) d\omega$$

$$= \int_{-\pi}^{\pi} \left\{ f_\theta(\omega)^{-1} \left(\frac{1}{2\pi} \int_{-\pi}^{\pi} f_\theta(\omega)^{-1} d\omega \right)^{-1} \right\}^2 f(\omega) d\omega$$

$$= 4\pi^2 \left(\int_{-\pi}^{\pi} \frac{1}{f_\theta(\omega)} d\omega \right)^{-2} \left(\int_{-\pi}^{\pi} \frac{f(\omega)}{f_\theta(\omega)^2} d\omega \right)$$

$$= 4\pi^2 D(f_\theta, f),$$

which shows that the disparity with $\alpha = -2$ and $\beta = -2$ is proportional to the interpolation error.

We investigate the fundamental properties of the disparity (2.4). For brevity, let $a(\theta)$ be

$$a(\theta) = \left(\int_{-\pi}^{\pi} f_\theta(\omega)^{\alpha+1} d\omega \right)^\beta, \quad \beta \in \mathbb{R}, \tag{2.9}$$

and the disparity (2.4) can be rewritten as

$$D_p(f_\theta, f) = \int_{-\pi}^{\pi} a(\theta) f_\theta(\omega)^\alpha f(\omega) d\omega, \quad \alpha \neq 0. \tag{2.10}$$

Denote by ∂ the partial derivative with respect to θ. For $1 \leq i \leq d$, let ∂_i be the partial derivative with respect to θ_i. We use the following expressions in our computation:

$$A_1(\theta) = \int_{-\pi}^{\pi} f_\theta(\lambda)^{\alpha+1} d\lambda, \qquad\qquad B_1(\theta)_i = f_\theta(\omega)^{\alpha-1} \partial_i f_\theta(\omega),$$

$$A_2(\theta)_i = \int_{-\pi}^{\pi} f_\theta(\lambda)^\alpha \partial_i f_\theta(\lambda) d\lambda, \qquad B_2(\theta) = f_\theta(\omega)^\alpha,$$

$$A_3(\theta)_{ij} = \int_{-\pi}^{\pi} f_\theta(\lambda)^{\alpha-1} \partial_i f_\theta(\lambda) \partial_j f_\theta(\lambda) d\lambda, \quad C_1(\theta) = \beta \left(\int_{-\pi}^{\pi} f_\theta(\lambda)^{\alpha+1} d\lambda \right)^{\beta-1}.$$

Suppose $f_{\theta_0} = f$ and f_θ is differentiable with respect to the parameter θ. Now let us determine the appropriate value of β from the basic requirements that the true parameter θ_0 must be a local optimizer of the disparity $D_p(f_\theta, f)$.

Theorem 2.1 *For the disparity (2.2), if either of the following cases*

(i) $\alpha = -1$ and θ is innovation free and
(ii) $\alpha \neq -1$ and $\beta = -\frac{\alpha}{\alpha+1}$ or

hold, then we have the following result:

$$\left. \partial_i D_p(f_\theta, f) \right|_{\theta=\theta_0} = 0, \quad \text{for all } 1 \leq i \leq d. \tag{2.11}$$

Proof First, let us consider the case (i). When $\alpha = -1$, Eq. (2.2) becomes

$$D_p(f_\theta, f) = C \int_{-\pi}^{\pi} f_\theta(\omega)^{-1} f(\omega) d\omega, \tag{2.12}$$

where C is a generic constant. On the other hand, from the general Kolmogorov formula (1.25) for the prediction error, the variance of the innovations of the model f_θ is expressed as

$$\sigma_\theta = 2\pi \exp\left\{ \frac{1}{2\pi} \int_{-\pi}^{\pi} \log f_\theta(\lambda) \right\}. \tag{2.13}$$

From the condition that θ is innovation free, σ_θ does not depend on θ. The partial derivative with respect to θ_i from both sides of (2.13) leads to the result

$$0 = 2\pi \exp\left\{ \frac{1}{2\pi} \int_{-\pi}^{\pi} \log f_\theta(\lambda) \right\} \partial_i \left(\frac{1}{2\pi} \int_{-\pi}^{\pi} \log f_\theta(\lambda) \right).$$

From the positivity of the exponential function, we have

$$\partial_i \left(\frac{1}{2\pi} \int_{-\pi}^{\pi} \log f_\theta(\lambda) \right) = \left(\frac{1}{2\pi} \int_{-\pi}^{\pi} f_\theta(\lambda)^{-1} \partial_i f_\theta(\lambda) \right) = 0. \tag{2.14}$$

By (2.12) and (2.14), we have

$$\left. \partial_i D_p(f_\theta, f) \right|_{\theta=\theta_0} = C \int_{-\pi}^{\pi} f_\theta(\omega)^{-2} \partial_i f_\theta(\omega) f(\omega) d\omega \Big|_{\theta=\theta_0} = 0,$$

which shows the conclusion (2.11).

We turn to the case (ii). When $\alpha \neq -1$, then

$$\left. \partial_i D_p(f_\theta, f_{\theta_0}) \right|_{\theta=\theta_0} = (\alpha + 1) C_1(\theta_0) A_1(\theta_0) A_2(\theta_0)_i + \beta^{-1} \alpha C_1(\theta_0) A_1(\theta_0) A_2(\theta_0)_i. \tag{2.15}$$

The conclusion follows from $\beta = -\alpha/(\alpha + 1)$. □

It can be seen that the true parameter θ_0 is a local optimizer of the disparity $D_p(f_\theta, f)$ under the conditions (i) or (ii) in Theorem 2.1. From now on, we consider our disparity (2.10) when these conditions hold and thus $\beta = -\alpha/(\alpha + 1)$. To clarify the notations, let us rewritten D_p in (2.4) as D, and Eqs. (2.9) and (2.10) become

$$a(\theta) = \left(\int_{-\pi}^{\pi} f_\theta(\omega)^{\alpha+1} d\omega \right)^{-\frac{\alpha}{\alpha+1}}, \qquad (2.16)$$

and the disparity (2.4) can be rewritten as

$$D(f_\theta, f) = \int_{-\pi}^{\pi} a(\theta) f_\theta(\omega)^\alpha f(\omega) d\omega, \quad \alpha \neq 0. \qquad (2.17)$$

Note that

$$C_1(\theta) = -\frac{\alpha}{\alpha + 1} \left(\int_{-\pi}^{\pi} f_\theta(\lambda)^{\alpha+1} d\lambda \right)^{-\frac{2\alpha+1}{\alpha+1}}.$$

Surprisingly, the disparity (2.17) has been discussed in Renyi (1961), (Csiszár 1975) and (Csiszár 1991) for the parameter estimation of i.i.d. random variables. The disparity (2.17) is also the optimal interpolation error if we substitute f_θ for the true spectral density f as we considered in Theorem 1.6 with $p = \alpha/(\alpha + 1)$. (See also Miamee and Pourahmadi 1988)

2.2 Fundamentals of the New Disparity

Let us consider the fundamental properties of the new disparity (2.17) with $a(\theta)$ in (2.16). Let $\mathscr{F}$ denote the set of all spectral densities with respect to the Lebesgue measure on $[-\pi, \pi]$. More specifically, we define $\mathscr{F}$ as

$$\mathscr{F} = \left\{ g : g(\omega) = \sigma^2 \left| \sum_{j=0}^{\infty} g_j \exp(-ij\omega) \right|^2 / (2\pi) \right\}.$$

Denote by $\mathscr{F}(\Theta)$ the set of spectral densities indexed by parameter θ.

Assumption 2.2 (i) The parameter space Θ is a compact subset of $\mathbb{R}^d$. θ_0 is the true parameter in the interior of the parameter space Θ and $f = f_{\theta_0} \in \mathscr{F}(\Theta)$.
  (ii) If $\theta_1 \neq \theta_2$, then $f_{\theta_1} \neq f_{\theta_2}$ on a set of positive Lebesgue measure.
  (iii) The parametric spectral density $f_\theta(\lambda)$ is three times continuously differentiable with respect to θ and the second derivative $\frac{\partial^2}{\partial \theta \partial \theta^\top} f_\theta(\lambda)$ is continuous in λ.

Under Assumption 2.2 (i) and (ii), we first examine the extreme value of the disparity (2.17) in the following. To prove it, we need Hölder's inequality for $p > 0$ ($p \neq 1$). Suppose q satisfies

$$\frac{1}{p} + \frac{1}{q} = 1.$$

Lemma 2.1 (Hewitt and Stromberg 1975) *Suppose $f \in L_p$, $g \in L_q$.*

(i) If $p > 1$, then

$$\|fg\|_1 \leq \|f\|_p \|g\|_q.$$

(ii) If $0 < p < 1$ and further suppose $f \in L_p^+$ and $g \in L_q^+$, then

$$\|fg\|_1 \geq \|f\|_p \|g\|_q.$$

The equality holds if and only if

$$|f|^p = C|g|^q, \quad almost\ everywhere.$$

Remark 2.1 If $0 < p < 1$, then $q < 0$ and vice versa. That is to say, (ii) is equivalent to the following condition

(ii)' if $p < 0$ and $f \in L_p^+$ and $g \in L_q^+$, then

$$\|fg\|_1 \geq \|f\|_p \|g\|_q.$$

Theorem 2.3 *Under Assumption 2.2 (i) and (ii), we have the following results:*

(i) If $\alpha > 0$, then θ_0 maximizes the disparity $D(f_\theta, f)$.
(ii) If $\alpha < 0$, then θ_0 minimizes the disparity $D(f_\theta, f)$.

Proof First, suppose $\alpha > 0$. The disparity (2.13) can be rewritten as

$$D(f_\theta, f_{\theta_0}) = \frac{\int_{-\pi}^{\pi} f_\theta(\omega)^\alpha f_{\theta_0}(\omega) d\omega}{(\int_{-\pi}^{\pi} f_\theta(\omega)^{\alpha+1} d\omega)^{\frac{\alpha}{\alpha+1}}}.$$

From Lemma 2.1, the numerator then satisfies

$$\int_{-\pi}^{\pi} f_\theta(\omega)^\alpha f_{\theta_0}(\omega) d\omega \leq \left(\int_{-\pi}^{\pi} \{f_\theta(\omega)^\alpha\}^{\frac{\alpha+1}{\alpha}} d\omega\right)^{\frac{\alpha}{\alpha+1}} \left(\int_{-\pi}^{\pi} f_{\theta_0}(\omega)^{\alpha+1} d\omega\right)^{\frac{1}{\alpha+1}}$$

$$= \left(\int_{-\pi}^{\pi} f_\theta(\omega)^{\alpha+1} d\omega\right)^{\frac{\alpha}{\alpha+1}} \left(\int_{-\pi}^{\pi} f_{\theta_0}(\omega)^{\alpha+1} d\omega\right)^{\frac{1}{\alpha+1}}.$$

Therefore,

$$D(f_\theta, f_{\theta_0}) \leq \left(\int_{-\pi}^{\pi} f_{\theta_0}(\omega)^{\alpha+1} d\omega\right)^{\frac{1}{\alpha+1}}.$$

The equality holds only when $f_\theta = f_{\theta_0}$ almost everywhere. From Assumption 2.2, the conclusion holds.

On the other hand, if $\alpha < 0$, then there are three cases (a) $-1 < \alpha < 0$, (b) $\alpha < -1$, and (c) $\alpha = -1$ must be considered. However, it is easy to see that both first two cases are corresponding to the case (ii) and (ii)' in Hölder's inequality since if $-1 < \alpha < 0$ then $(\alpha + 1)/\alpha < 0$ and if $\alpha < -1$ then $0 < (\alpha + 1)/\alpha < 1$. As a result, we obtain

$$D(f_\theta, f_{\theta_0}) \geq \left(\int_{-\pi}^\pi f_{\theta_0}(\omega)^{\alpha+1} d\omega \right)^{\frac{1}{\alpha+1}},$$

with a minimum from Assumption 2.2. For the case (c), the disparity is corresponding to the predictor error. There is a lower bound for the disparity. (See Proposition 10.8.1 in Brockwell and Davis 1991.)   □

Last, let us show a stronger result for the disparity (2.7) under Assumption 2.2. To see the result, we provide the following Lemma as a preparation, which is a generalization of the Cauchy–Bunyakovsky inequality, first used in Grenander and Rosenblatt (1957) in the context of time series analysis and then the paper by Kholevo (1969) later on.

Lemma 2.2 (Grenander and Rosenblatt 1957, Kholevo 1969) *Let $A(\omega)$, $B(\omega)$ be $r \times s$ matrix-valued functions, and $g(\omega)$ a positive function on $\omega \in [-\pi, \pi]$. If*

$$\left\{ \int_{-\pi}^\pi B(\omega) B(\omega)^{\mathsf{T}} g(\omega)^{-1} d\omega \right\}^{-1}$$

exists, the following inequality:

$$\int_{-\pi}^\pi A(\omega) A(\omega)^{\mathsf{T}} g(\omega) d\omega$$
$$\geq \left\{ \int_{-\pi}^\pi A(\omega) B(\omega)^{\mathsf{T}} d\omega \right\} \left\{ \int_{-\pi}^\pi B(\omega) B(\omega)^{\mathsf{T}} g(\omega)^{-1} d\omega \right\}^{-1} \left\{ \int_{-\pi}^\pi A(\omega) B(\omega)^{\mathsf{T}} d\omega \right\}^{\mathsf{T}}$$

$$\tag{2.18}$$

holds. In (2.18), the equality holds if and only if there exists a constant matrix C such that
$$g(\omega) A(\omega) + C B(\omega) = O, \quad \text{almost everywhere } \omega \in [-\pi, \pi]. \tag{2.19}$$

Theorem 2.4 *Under Assumption 2.2, we have the following results:*

(i) *If $\alpha > 0$, then the disparity $D(f_\theta, f)$ is convex upward with respect to θ.*
(ii) *If $\alpha < 0$, then the disparity $D(f_\theta, f)$ is convex downward with respect to θ.*

Proof Suppose $\alpha \neq -1$. Then

$$\partial_i D(f_\theta, g) = (\alpha + 1) C_1(\theta) \left\{ A_1(\theta) \int_{-\pi}^\pi B_1(\theta)_i g(\omega) d\omega - A_2(\theta) \int_{-\pi}^\pi B_2(\theta)_i g(\omega) d\omega \right\}.$$

If $g = f_\theta$, then it is easy to see that $\partial_i D(f_\theta, g) = 0$ for any $i = 1, \ldots, d$. Considering twice derivative of $D(f_\theta, g)$, we have

$$\partial_i \partial_j D(f_\theta, g) = (\alpha + 1)C_1(\theta)\Big(A_1(\theta)A_3(\theta)_{ij} - A_2(\theta)_i A_2(\theta)_j\Big)$$

if $g = f_\theta$. Let us set the expressions $A(\omega)$, $B(\omega)$ and $g(\omega)$ in Lemma 2.2 as follows:

$$A(\omega) = f_\theta(\omega)^{\alpha/2},$$
$$B(\omega) = f_\theta(\omega)^{\alpha/2}\partial_i f_\theta(\omega),$$
$$g(\omega) = f_\theta(\omega).$$

Then we can see that the matrix $A_1(\theta)A_3(\theta)_{ij} - A_2(\theta)_i A_2(\theta)_j$ is positive definite. Therefore, the conclusion of the convexity of the disparity holds. The convexity in the case of $\alpha = -1$ is also easy to show. □

From Theorem 2.3, it is shown that the true parameter $\theta_0 \in \Theta$ is an optimizer of the criterion $D(f_\theta, f)$. The results are different at the two sides of $\alpha = 0$. To keep uniformity of the context, we suppose $\alpha < 0$ hereafter. Accordingly, we can define the functional T as

$$D(f_{T(g)}, g) = \min_{t \in \Theta} D(f_t, g), \quad \text{for every} g \in \mathcal{F}. \tag{2.20}$$

Thus, if the model is specified, that is, $f \in \mathcal{F}(\Theta)$, then $\theta_0 = T(f)$.

2.3 Parameter Estimation Based on Disparity

In this section, we elucidate the parameter estimation based on the disparity (2.17). We first provide the fundamental properties of estimation based on the disparity. We study the minimum contrast estimators for stationary processes with finite variance innovations and infinite variance innovations in the next two subsections, respectively.

Theorem 2.5 *Under Assumption 2.2, we have the following results:*

(i) *For every $f \in \mathcal{F}$, there exists a value $T(f) \in \Theta$ satisfying (2.20).*
(ii) *If $T(f)$ is unique and if $f_n \xrightarrow{L^2} f$, then $T(f_n) \to T(f)$ as $n \to \infty$.*
(iii) *$T(f_\theta) = \theta$ for every $\theta \in \Theta$.*

Proof (i) Define $h(\theta)$ as $h(\theta) = D(f_\theta, f)$. If the continuity of $h(\theta)$ in $\theta \in \Theta$ is shown, then the existence of $T(f)$ follows the compactness of Θ. From the proof in Theorem 2.3,

$$h(\theta) \leq \left(\int_{-\pi}^{\pi} f(\omega)^{\alpha+1} d\omega\right)^{\frac{1}{\alpha+1}} \leq C.$$

By Lebesgue's dominated convergence theorem,

$$|h(\theta_n) - h(\theta)| \le \left| \int_{-\pi}^{\pi} \left(a(\theta_n) f_{\theta_n}(\omega)^{\alpha} - a(\theta) f_{\theta}(\omega)^{\alpha} \right) f(\omega) d\omega \right| \to 0$$

for any convergence sequence $\{\theta_n \in \Theta; \theta_n \to \theta\}$, which shows the continuity of $h(\theta)$.

(ii) Similarly, suppose $h_n(\theta) = D(f_{\theta}, f_n)$. Then

$$\lim_{n \to \infty} \sup_{\theta \in \Theta} |h_n(\theta) - h(\theta)|$$

$$= \lim_{n \to \infty} \sup_{\theta \in \Theta} \left| \int_{-\pi}^{\pi} (a(\theta) f_{\theta}^{\alpha}(\omega))(f_n(\omega) - f(\omega)) d\omega \right|$$

$$\le \lim_{n \to \infty} \sup_{\theta \in \Theta} \left| \int_{-\pi}^{\pi} (a(\theta) f_{\theta}^{\alpha}(\omega))^2 d\omega \int_{-\pi}^{\pi} (f_n(\omega) - f(\omega))^2 d\omega \right|^{1/2}$$

$$\le C \lim_{n \to \infty} \sup_{\theta \in \Theta} \left| \int_{-\pi}^{\pi} (f_n(\omega) - f(\omega))^2 d\omega \right|^{1/2}$$

$$= 0, \tag{2.21}$$

by $f_n \xrightarrow{L^2} f$. From (2.21), we have

$$|h_n(T(f_n)) - h(T(f_n))| \to 0.$$

In addition, the uniform convergence (2.21) implies that

$$\left| \min_{\theta \in \Theta} h_n(\theta) - \min_{\theta \in \Theta} h(\theta) \right| \to 0.$$

In other words, we have

$$|h_n(T(f_n)) - h(T(f))| \to 0.$$

By the triangle inequality, it holds that

$$|h(T(f_n)) - h(T(f))| \le |h(T(f_n)) - h_n(T(f_n))| + |h_n(T(f_n)) - h(T(f))| \to 0,$$

and therefore
$$h(T(f_n)) \to h(T(f)).$$

The continuity of h and the uniqueness of $T(f)$ show the conclusion, that is,

$$T(f_n) \to T(f).$$

(iii) From the definition (2.20), $T(f_\theta)$ is the minimizer of $D(f_t, f_\theta)$ for $t \in \Theta$. The conclusion is a direct result of Theorem 2.3 (ii), which we have shown before. □

Now we move on to the Hessian matrix of the estimation procedure (2.20). This can be regarded as a continuation of Theorem 2.4 from the view of estimation.

Theorem 2.6 *Under Assumptions 2.2, we have*

$$T(f_n) = \theta_0 - \int_{-\pi}^{\pi} \rho(\omega)(f_n(\omega) - f(\omega))d\omega + O(\|f_n - f\|^2),$$

for every spectral density sequence $\{f_n\}$ satisfying $f_n \xrightarrow{L^2} f$, where

$$\rho(\omega) = \left(A_1(\theta_0)A_3(\theta_0) - A_2(\theta_0)A_2(\theta_0)^\mathsf{T}\right)^{-1}\left(A_1(\theta_0)B_1(\theta_0) - A_2(\theta_0)B_2(\theta_0)\right).$$

Proof Note that $\theta_0 = T(f)$. From Theorem 2.1, for all $1 \le i \le d$, we have

$$\partial_i D(f_\theta, f_n)\Big|_{\theta=T(f_n)} = 0,$$

$$\partial_i D(f_\theta, f)\Big|_{\theta=\theta_0} = 0.$$

Then there exists a $\theta^* \in \mathbb{R}^d$ on the line joining $T(f_n)$ and θ_0 such that

$$T(f_n) - \theta_0 = \left\{(\alpha + 1)C_1(\theta^*)\left(A_1(\theta^*)A_3(\theta^*) - A_2(\theta^*)A_2(\theta^*)^\mathsf{T}\right)\right\}^{-1}$$
$$\int_{-\pi}^{\pi}\left(A_1(\theta_0)B_1(\theta_0) - A_2(\theta_0)B_2(\theta_0)\right)(f_n - f)d\omega + O(\|f_n - f\|^2).$$

$$(2.22)$$

Note that from Theorem 2.5 (ii), we have the following inequality:

$$\left\|(\alpha + 1)C_1(\theta^*)\left(A_1(\theta^*)A_3(\theta^*) - A_2(\theta^*)A_2(\theta^*)^\mathsf{T}\right)\right.$$
$$\left. - \left((\alpha + 1)C_1(\theta_0)(A_1(\theta_0)A_3(\theta_0) - A_2(\theta_0)A_2(\theta_0)^\mathsf{T})\right)\right\| \le C|f_n - f|^2,$$

$$(2.23)$$

where $C > 0$ is a generic constant. The conclusion follows from (2.22) and (2.23). □

2.3.1 Finite Variance Innovations Case

In this subsection, we investigate asymptotic behavior of the parameter estimation based on disparity. Suppose $\{X(t) : t \in \mathbb{Z}\}$ is a zero-mean stationary process and

$$X(t) = \sum_{j=0}^{\infty} g_j \varepsilon(t - j), \quad t \in \mathbb{Z},$$

where $\{\varepsilon(t)\}$ is a stationary innovation process with finite fourth-order moment $E\varepsilon(t)^4 < \infty$ and satisfies $E[\,\varepsilon(t)\,] = 0$ and $\mathrm{Var}[\,\varepsilon(t)\,] = \sigma^2$ with $\sigma^2 > 0$. We impose the following regularity conditions.

Assumption 2.7 For all $|z| \le 1$, there exist $C < \infty$ and $\delta > 0$ such that

(i) $\sum_{j=0}^{\infty}(1 + j^2)|g_j| \le C$;

(ii) $\left|\sum_{j=0}^{\infty} g_j z^j\right| \ge \delta$;

(iii) $\sum_{t_1,t_2,t_3=-\infty}^{\infty}|Q_\varepsilon(t_1, t_2, t_3)| < \infty$, where $Q_\varepsilon(t_1, t_2, t_3)$ is the fourth-order cumulant of $\varepsilon(t)$, $\varepsilon(t + t_1)$, $\varepsilon(t + t_2)$ and $\varepsilon(t + t_3)$.

Assumption 2.7 (iii) guarantees the existence of a fourth-order spectral density. The fourth-order spectral density $\tilde{Q}_\varepsilon(\omega_1, \omega_2, \omega_3)$ is

$$\tilde{Q}_\varepsilon(\omega_1, \omega_2, \omega_3) = \left(\frac{1}{2\pi}\right)^3 \sum_{t_1,t_2,t_3=-\infty}^{\infty} Q_\varepsilon(t_1, t_2, t_3)e^{-i(\omega_1 t_1 + \omega_2 t_2 + \omega_3 t_3)}.$$

Denote by $(X(1), \ldots, X(n))$ the observations from the process $\{X(t)\}$. Let $I_{n,X}(\omega)$ be the periodogram of observations, that is,

$$I_{n,X}(\omega) = \frac{1}{2\pi n}\left|\sum_{t=1}^{n} X(t)e^{it\omega}\right|^2, \quad -\pi \le \omega \le \pi.$$

Under Assumption 2.2, we can define the estimator $\hat{\theta}_n$ based on (2.20) as

$$\hat{\theta}_n = \arg\min_{\theta \in \Theta} D(f_\theta, I_{n,X}). \tag{2.24}$$

Now we state the regularity conditions for the parameter estimation by $\hat{\theta}_n$. Let $\mathscr{B}(t)$ denote the σ-field generated by $\varepsilon(s)$ $(-\infty < s \le t)$.

Assumption 2.8

(i) For each nonnegative integer m and $\eta_1 > 0$,

$$\mathrm{Var}[E\big(\varepsilon(t)\varepsilon(t + m)|\mathscr{B}(t - \tau)\big)] = O(\tau^{-2-\eta_1})$$

uniformly in t.

(ii) For any $\eta_2 > 0$,

$$E|E\{\varepsilon(t_1)\varepsilon(t_2)\varepsilon(t_3)\varepsilon(t_4)|\mathscr{B}(t_1 - \tau)\} - E\big(\varepsilon(t_1)\varepsilon(t_2)\varepsilon(t_3)\varepsilon(t_4)\big)| = O(\tau^{-1-\eta_2}),$$

uniformly in t_1, where $t_1 \leq t_2 \leq t_3 \leq t_4$.

(iii) For any $\eta_3 > 0$ and for any fixed integer $L \geq 0$, there exists $B_{\eta_3} > 0$ such that

$$E[T(n, s)^2 \mathbb{1}\{T(n, s) > B_{\eta_3}\}] < \eta_3$$

uniformly in n and s, where

$$T(n, s) = \left[\frac{1}{n}\sum_{r=0}^{L}\left\{\sum_{t=1}^{n} \varepsilon(t + s)\varepsilon(t + s + r) - \sigma^2\delta(0, r)\right\}^2\right]^{1/2}.$$

Theorem 2.9 *Suppose Assumptions 2.2, 2.7 and 2.8 hold. As for the spectral density $f \in \mathscr{F}(\Theta)$, the estimator $\hat{\theta}_n$ defined by (2.24) has the following asymptotic properties:*

(i) *$\hat{\theta}_n$ converges to θ_0 in probability;*

(ii) *The distribution of $\sqrt{n}(\hat{\theta}_n - \theta_0)$ is asymptotically normal with mean 0 and covariance matrix $H(\theta_0)^{-1} V(\theta_0) H(\theta_0)^{-1}$, where*

$$H(\theta_0) = \left(\int_{-\pi}^{\pi} f_{\theta_0}(\omega)^\alpha \partial f_{\theta_0}(\omega) d\omega\right)\left(\int_{-\pi}^{\pi} f_{\theta_0}(\omega)^\alpha \partial f_{\theta_0}(\omega) d\omega\right)^{\mathrm{T}}$$
$$- \int_{-\pi}^{\pi} f_{\theta_0}(\omega)^{\alpha+1} d\omega \int_{-\pi}^{\pi} f_{\theta_0}(\omega)^{\alpha-1}\big(\partial f_{\theta_0}(\omega)\big)\big(\partial f_{\theta_0}(\omega)\big)^{\mathrm{T}} d\omega,$$

$$V(\theta_0) = 4\pi \int_{-\pi}^{\pi}\left(f_{\theta_0}(\omega)^\alpha \partial f_{\theta_0}(\omega)\int_{-\pi}^{\pi} f_{\theta_0}(\lambda)^{\alpha+1} d\lambda\right.$$
$$- f_{\theta_0}(\omega)^{\alpha+1}\int_{-\pi}^{\pi} f_{\theta_0}(\lambda)^\alpha \partial f_{\theta_0}(\lambda) d\lambda\Big)$$
$$\times\left(f_{\theta_0}(\omega)^\alpha \partial f_{\theta_0}(\omega)\int_{-\pi}^{\pi} f_{\theta_0}(\lambda)^{\alpha+1} d\lambda\right.$$
$$- f_{\theta_0}(\omega)^{\alpha+1}\int_{-\pi}^{\pi} f_{\theta_0}(\lambda)^\alpha \partial f_{\theta_0}(\lambda) d\lambda\Big)^{\mathrm{T}} d\omega$$
$$+ 2\pi \iint_{-\pi}^{\pi}\left(f_{\theta_0}(\omega_1)^{\alpha-1} \partial f_{\theta_0}(\omega_1)\int_{-\pi}^{\pi} f_{\theta_0}(\lambda)^{\alpha+1} d\lambda\right.$$
$$- f_{\theta_0}(\omega_1)^\alpha\int_{-\pi}^{\pi} f_{\theta_0}(\lambda)^\alpha \partial f_{\theta_0}(\lambda) d\lambda\Big)$$
$$\times\left(f_{\theta_0}(\omega_2)^{\alpha-1} \partial f_{\theta_0}(\omega_2)\int_{-\pi}^{\pi} f_{\theta_0}(\lambda)^{\alpha+1} d\lambda\right.$$
$$- f_{\theta_0}(\omega_2)^\alpha\int_{-\pi}^{\pi} f_{\theta_0}(\lambda)^\alpha \partial f_{\theta_0}(\lambda) d\lambda\Big)^{\mathrm{T}}$$
$$\times \tilde{Q}_X(-\omega_1, \omega_2, -\omega_2) d\omega_1 d\omega_2. \tag{2.25}$$

Here, $\tilde{Q}_X(\omega_1, \omega_2, \omega_3) = A(\omega_1)A(\omega_2)A(\omega_3)A(-\omega_1 - \omega_2 - \omega_3)\tilde{Q}_\varepsilon(\omega_1,$
$\omega_2, \omega_3)$ *and* $A(\omega) = \sum_{j=0}^{\infty} g_j \exp(ij\omega)$.

Proof In view of (2.21), it is equivalent to consider $\hat{\theta}_n$ satisfies

$$\partial D(f_\theta, I_{n,X})\Big|_{\theta=\hat{\theta}_n} = 0.$$

The result that $\hat{\theta}_n \to \theta_0$ in probability follows that for any $\theta \in \Theta$ compact,

$$\partial D(f_\theta, I_{n,X}) \to \partial D(f_\theta, f_{\theta_0}) \quad \text{in probability},$$

which is guaranteed by Lemma 3.3A in Hosoya and Taniguchi (1982). Differentiating
the disparity (2.13) with respect to θ, then we have

$$\partial D(f_\theta, I_{n,X}) = C_1(\theta) \int_{-\pi}^{\pi} (A_1(\theta)B_1(\theta) - A_2(\theta)B_2(\theta))I_{n,X}(\omega)d\omega.$$

From $\partial D(f_\theta, f_{\theta_0})\Big|_{\theta=\theta_0} = 0$ as in (2.15), the asymptotic normality of the estimator
follows from Assumption 2.7, that is,

$$\partial D(f_\theta, I_{n,X})\Big|_{\theta=\theta_0} =$$
$$C_1(\theta_0) \int_{-\pi}^{\pi} (A_1(\theta_0)B_1(\theta_0) - A_2(\theta_0)B_2(\theta_0))(I_{n,X}(\omega) - f_{\theta_0}(\omega))d\omega$$

converges in distribution to a normal distribution with mean 0 and variance matrix
$C_1(\theta_0)^2 V(\theta_0)$. Using $\partial D(f_\theta, f_{\theta_0})\Big|_{\theta=\theta_0} = 0$ again, we see that

$$\left|\partial^2 D(f_\theta, I_{n,X}) - C_1(\theta)\partial\left(\int_{-\pi}^{\pi} (A_1(\theta)B_1(\theta) - A_2(\theta)B_2(\theta))I_{n,X}(\omega)d\omega\right)\Big|_{\theta=\theta_0}\right| \to 0$$

in probability. We also have

$$\int_{-\pi}^{\pi} B_1(\theta)f_\theta(\omega)d\omega \, \partial A_1(\theta) - \int_{-\pi}^{\pi} A_2(\theta)f_\theta(\omega)d\omega \, \partial B_2(\theta) = A_2(\theta)A_2(\theta)^{\mathrm{T}}$$
$$\int_{-\pi}^{\pi} A_1(\theta)f_\theta(\omega)d\omega \, \partial B_1(\theta) - \int_{-\pi}^{\pi} B_2(\theta)f_\theta(\omega)d\omega \, \partial A_2(\theta) = -A_1(\theta)A_3(\theta),$$

and therefore

$$C_1(\theta)\partial\left(\int_{-\pi}^{\pi} (A_1(\theta)B_1(\theta) - A_2(\theta)B_2(\theta))I_{n,X}(\omega)d\omega\right)\Big|_{\theta=\theta_0} \to C_1(\theta_0)H(\theta_0).$$

in probability. As a result, we obtain

$$\partial^2 D(f_\theta, I_{n,X})\Big|_{\theta=\theta_0} \to C_1(\theta_0)H(\theta_0).$$

in probability. Canceling $C_1(\theta_0)$, the desirable result is obtained.   □

The asymptotic variance (2.25) of the estimator $\hat{\theta}_n$ seems extremely complex. Sometimes we are not interested in all disparities (2.17) for different α but some. Especially, when $\alpha = -1$, the disparity corresponds to the prediction error.

Example 2.3 Let us give the result for the Gaussian stationary process under the special case $\alpha = -1$. Suppose the model f_θ is innovation free. From Theorem 1.3, it holds that

$$\sigma^2 = 2\pi \exp\left(\frac{1}{2\pi}\int_{-\pi}^{\pi} \log f_\theta(\lambda)d\lambda\right).$$

Note that it holds that

$$\int_{-\pi}^{\pi} f_\theta(\lambda)^{-1}\partial f_\theta(\lambda)d\lambda = \partial \int_{-\pi}^{\pi} \log f_\theta(\lambda)d\lambda = 0. \tag{2.26}$$

When $\alpha = -1$, by (2.26), we have

$$H(\theta_0) = 2\pi \int_{-\pi}^{\pi} f_{\theta_0}(\omega)^{-2}\Big(\partial f_{\theta_0}(\omega)\Big)\Big(\partial f_{\theta_0}(\omega)\Big)^{\mathrm{T}}d\omega.$$

Note that $\tilde{Q}_X(-\omega_1, \omega_2, -\omega_2) = 0$ for the Gaussian process. By (2.26) again, we have

$$V(\theta_0) = 16\pi^3 \int_{-\pi}^{\pi} f_{\theta_0}(\omega)^{-2}\Big(\partial f_{\theta_0}(\omega)\Big)\Big(\partial f_{\theta_0}(\omega)\Big)^{\mathrm{T}}d\omega.$$

Therefore, the asymptotic covariance matrix for $\sqrt{n}(\hat{\theta}_n - \theta_0)$ is

$$H(\theta_0)^{-1}V(\theta_0)H(\theta_0)^{-1} = 4\pi\left(\int_{-\pi}^{\pi} f_{\theta_0}(\omega)^{-2}\Big(\partial f_{\theta_0}(\omega)\Big)\Big(\partial f_{\theta_0}(\omega)\Big)^{\mathrm{T}}d\omega\right)^{-1}. \tag{2.27}$$

Generally, the inverse of the right-hand side of (2.27) is called *the Gaussian Fisher information matrix* in time series analysis. Let us denote it by $\mathscr{F}(\theta_0)$, i.e.,

$$\mathscr{F}(\theta_0) = \frac{1}{4\pi}\int_{-\pi}^{\pi} f_{\theta_0}^{-2}(\lambda)\partial f_{\theta_0}(\lambda)\partial f_{\theta_0}(\lambda)^{\mathrm{T}}d\lambda. \tag{2.28}$$

An estimator $\hat{\theta}_n$ is said to be *Gaussian asymptotically efficient* if $\sqrt{n}(\hat{\theta}_n - \theta_0) \xrightarrow{\mathscr{L}} \mathscr{N}(0, \mathscr{F}(\theta_0)^{-1})$.

2.3.2 Infinite Variance Innovations Case

In this subsection, we consider the linear processes with infinite variance innovations. Suppose $\{X(t); t = 1, 2, \ldots\}$ is a zero-mean stationary process generated by

$$X(t) = \sum_{j=0}^{\infty} g_j \varepsilon(t - j), \quad t = 1, 2, \ldots,$$

where i.i.d. symmetric innovation process $\{\varepsilon(t)\}$ satisfies the following assumptions.

Assumption 2.10 For some $k > 0$, $\delta = 1 \wedge k$ and positive sequence a_n satisfying $a_n \uparrow \infty$, the coefficient g_j and the innovation process $\{\varepsilon(t)\}$ have the following properties:

(i) $\sum_{j=0}^{\infty} |j| \|g_j\|^{\delta} < \infty$;
(ii) $E|\varepsilon(t)|^k < \infty$;
(iii) as $n \to \infty$, $n/a_n^{2\delta} \to 0$;
(iv) $\lim_{x \to 0} \lim \sup_{n \to \infty} P\left(a_n^{-2} \sum_{t=1}^{n} \varepsilon(t)^2 \le x\right) = 0$;
(v) For some $0 < q < 2$, the distribution of $\varepsilon(t)$ is in the domain of normal attraction of a symmetric q-stable random variable Y.

For Assumption 2.10, note that the positive sequence a_n can be specified from (v) by choosing $a_n = n^{1/q}$ for $n \ge 1$. (See (Feller 1968), (Bingham et al. 1987).)

An issue concerning the infinite variance innovations is the periodogramWhittle likelihood $I_{n,X}(\omega)$ is not well defined in this case. For this type of stationary process, the self-normalized periodogram $\tilde{I}_{n,X}(\omega) = |\sum_{t=1}^{n} X(t)e^{it\omega}|^2 / \sum_{t=1}^{n} X(t)^2$ is substituted for the periodogram $I_{n,X}(\omega)$. Let us define the power transfer function $f(\omega)$ by

$$f(\omega) = \left|\sum_{j=0}^{\infty} g_j e^{ij\omega}\right|^2, \quad \omega \in [-\pi, \pi].$$

Again it is possible to formulate the approach to estimate the parameter by minimizing the disparity (2.17). That is, we fit a parametric model f_θ to the self-normalized periodogram $\tilde{I}_{n,X}(\omega)$:

$$\hat{\theta}_n = \arg\min_{\theta \in \Theta} D(f_\theta, \tilde{I}_{n,X}). \tag{2.29}$$

For the case of infinite variance innovations, we introduce the scale constant C_q appearing in the asymptotic distribution, i.e.,

$$C_q = \begin{cases} \dfrac{1-q}{\Gamma(2-q)\cos(\pi q/2)}, & \text{if } q \ne 1, \\ \dfrac{2}{\pi}, & \text{if } q = 1. \end{cases}$$

Theorem 2.11 *Suppose Assumptions 2.2 and 2.10 hold. As for the power transfer function $f \in \mathcal{F}(\Theta)$, the estimator $\hat{\theta}_n$ defined by (2.29) has the following asymptotic properties:*

(i) $\hat{\theta}_n$ *converges to θ_0 in probability;*
(ii) *It holds that*

$$\left(\frac{n}{\log n}\right)^{1/q}(\hat{\theta}_n - \theta_0) \to 4\pi H^{-1}(\theta_0) \sum_{k=1}^{\infty} \frac{Y_k}{Y_0} V_k(\theta_0) \qquad (2.30)$$

in law, where $H(\theta_0)$ is the same as in Theorem 2.9,

$$V_k(\theta_0) = \left(\int_{-\pi}^{\pi} f_{\theta_0}(\omega)^\alpha \partial f_{\theta_0}(\omega) d\omega\right)\left(\int_{-\pi}^{\pi} f_{\theta_0}(\omega)^{\alpha+1} \cos(k\omega) d\omega\right)$$
$$- \left(\int_{-\pi}^{\pi} f_{\theta_0}(\omega)^{\alpha+1} d\omega\right)\left(\int_{-\pi}^{\pi} f_{\theta_0}(\omega)^\alpha \partial f_{\theta_0}(\omega) \cos(k\omega) d\omega\right),$$

and $\{Y_k\}_{k=0,1,\ldots}$ are mutually independent random variables. Y_0 is $q/2$-stable with scale $C_{q/2}^{-2/q}$ and Y_k $(k \geq 1)$ is q-stable with scale $C_q^{-1/q}$.

Proof For the proof of Theorem 2.11, we only need to change $g(\lambda, \beta)^{-1}$ in Mikosch et al. (1995) into $a(\theta) f_{\theta_0}(\omega)^\alpha$ to show the statements. □

Theorem 2.11 shows that the asymptotic distribution (2.30) in the case of infinite variance innovations is quite different from the normal distribution (2.25) in the case of finite variance innovations. Our minimum contrast estimator based on the disparity, however, has consistency and well-defined asymptotic distribution not only in the case of the finite variance innovations but also the infinite variance innovations.

2.4 Efficiency and Robustness

In this section, we discuss the asymptotic efficiency and robustness of the estimator $\hat{\theta}_n$ in (2.24). Especially, we discuss the asymptotic efficient estimator in the new class of disparities and illustrate it by some examples. On the other hand, the estimator $\hat{\theta}_n$ is robust in their asymptotic distribution in the sense that it does not depend on the fourth-order cumulant under some suitable conditions and robust from its intrinsic feature against randomly missing observations from time series.

2.4.1 Robustness Against the Fourth-Order Cumulant

The estimator $\hat{\theta}_n$ is said to be robust against the fourth-order cumulant if the asymptotic variance of $\hat{\theta}_n$ does not depend on $\tilde{Q}_X(-\omega_1, \omega_2, -\omega_2)$. The estimator $\hat{\theta}_n$ is robust against the fourth-order cumulant under the following assumption.

Assumption 2.12 For the innovation process $\{\varepsilon(t)\}$, suppose the fourth-order cumulant $\mathrm{cum}\big(\varepsilon(t_1), \varepsilon(t_2), \varepsilon(t_3), \varepsilon(t_4)\big)$ satisfies

$$\mathrm{cum}\big(\varepsilon(t_1), \varepsilon(t_2), \varepsilon(t_3), \varepsilon(t_4)\big) = \begin{cases} \kappa_4 & \text{if } t_1 = t_2 = t_3 = t_4, \\ 0 & \text{otherwise.} \end{cases}$$

If Assumption 2.12 holds, the fourth-order spectral density $\tilde{Q}_X(\omega_1, \omega_2, \omega_3)$ of $\{X(t)\}$ becomes

$$\tilde{Q}_X(\omega_1, \omega_2, \omega_3) = (2\pi)^{-3}\kappa_4 A(\omega_1)A(\omega_2)A(\omega_3)A(-\omega_1 - \omega_2 - \omega_3). \qquad (2.31)$$

Thus, we obtain the following theorem.

Theorem 2.13 *Suppose Assumptions 2.2, 2.7, 2.8 and 2.12 hold. As for the spectral density $f \in \mathscr{F}(\Theta)$, The distribution of $\sqrt{n}(\hat{\theta}_n - \theta_0)$ is asymptotically Gaussian with mean 0 and variance $H(\theta_0)^{-1}\tilde{V}(\theta_0)H(\theta_0)^{-1}$, where*

$$\begin{aligned}
\tilde{V}(\theta_0) = 4\pi \int_{-\pi}^{\pi} &\left(f_{\theta_0}(\omega)^\alpha \partial f_{\theta_0}(\omega) \int_{-\pi}^{\pi} f_{\theta_0}(\lambda)^{\alpha+1} d\lambda \right. \\
&\left. - f_{\theta_0}(\omega)^{\alpha+1} \int_{-\pi}^{\pi} f_{\theta_0}(\lambda)^\alpha \partial f_{\theta_0}(\lambda) d\lambda \right) \\
\times &\left(f_{\theta_0}(\omega)^\alpha \partial f_{\theta_0}(\omega) \int_{-\pi}^{\pi} f_{\theta_0}(\lambda)^{\alpha+1} d\lambda \right. \\
&\left. - f_{\theta_0}(\omega)^{\alpha+1} \int_{-\pi}^{\pi} f_{\theta_0}(\lambda)^\alpha \partial f_{\theta_0}(\lambda) d\lambda \right)^{\mathsf{T}} d\omega. \qquad (2.32)
\end{aligned}$$

Proof By the expression (2.31), we have

$$\begin{aligned}
&\tilde{Q}_X(-\omega_1, \omega_2, -\omega_2) = \\
&\frac{\kappa_4}{2\pi\sigma^4}\left(\frac{\sigma^2}{2\pi}\right)^2 A(-\omega_1)A(\omega_2)A(-\omega_2)A(\omega_1) = \frac{\kappa_4}{2\pi\sigma^4} f_{\theta_0}(\omega_1) f_{\theta_0}(\omega_2). \qquad (2.33)
\end{aligned}$$

By (2.33), the second term in (2.25) can be evaluated by

$$2\pi \iint_{-\pi}^{\pi} \left(f_{\theta_0}(\omega_1)^{\alpha-1} \partial f_{\theta_0}(\omega_1) \int_{-\pi}^{\pi} f_{\theta_0}(\lambda)^{\alpha+1} d\lambda \right.$$
$$\left. - f_{\theta_0}(\omega_1)^{\alpha} \int_{-\pi}^{\pi} f_{\theta_0}(\lambda)^{\alpha} \partial f_{\theta_0}(\lambda) d\lambda \right)$$
$$\times \left(f_{\theta_0}(\omega_2)^{\alpha-1} \partial f_{\theta_0}(\omega_2) \int_{-\pi}^{\pi} f_{\theta_0}(\lambda)^{\alpha+1} d\lambda \right.$$
$$\left. - f_{\theta_0}(\omega_2)^{\alpha} \int_{-\pi}^{\pi} f_{\theta_0}(\lambda)^{\alpha} \partial f_{\theta_0}(\lambda) d\lambda \right)^{\mathrm{T}} \tilde{Q}_X(-\omega_1, \omega_2, -\omega_2) d\omega_1 d\omega_2$$

$$= \frac{\kappa_4}{\sigma^4} \left(\int_{-\pi}^{\pi} f_{\theta_0}(\omega_1)^{\alpha} \partial f_{\theta_0}(\omega_1) d\omega_1 \int_{-\pi}^{\pi} f_{\theta_0}(\lambda)^{\alpha+1} d\lambda \right.$$
$$\left. - \int_{-\pi}^{\pi} f_{\theta_0}(\omega_1)^{\alpha+1} d\omega_1 \int_{-\pi}^{\pi} f_{\theta_0}(\lambda)^{\alpha} \partial f_{\theta_0}(\lambda) d\lambda \right)$$
$$\times \left(\int_{-\pi}^{\pi} f_{\theta_0}(\omega_2)^{\alpha} \partial f_{\theta_0}(\omega_2) d\omega_2 \int_{-\pi}^{\pi} f_{\theta_0}(\lambda)^{\alpha+1} d\lambda \right.$$
$$\left. - \int_{-\pi}^{\pi} f_{\theta_0}(\omega_2)^{\alpha+1} d\omega_2 \int_{-\pi}^{\pi} f_{\theta_0}(\lambda)^{\alpha} \partial f_{\theta_0}(\lambda) d\lambda \right)^{\mathrm{T}}$$

$$= O_d,$$

where O_d denotes the $d \times d$ zero matrix. This shows why the second term in Eq. (2.25) vanishes when we take the prediction and interpolation error as a disparity. $\square$

Assumptions 2.12 seems strong. However, for example, the Gaussian process always satisfies Assumption 2.12. In practice, modeling a process is usually up to second order. Making an assumption on simultaneous fourth-order cumulants covers a sufficiently large family of models.

Let us compare Eq. (2.32) with Eq. (2.25) in Theorem 2.9. The term with the fourth-order spectral density $\tilde{Q}_X(\omega_1, \omega_2, \omega_3)$ of $\{X(t)\}$ vanishes. This fact is well known for the case $\alpha = -1$, i.e., the Whittle likelihood estimator is robust against the fourth-order cumulant. We have shown that the robustness against the fourth-order cumulant also holds for any $\alpha \in \mathbb{R} \setminus \{0\}$.

2.4.2 Asymptotic Efficiency

As shown in (2.28), the variance of the estimator $\hat{\theta}_n$ minimizing prediction error is asymptotically $\mathscr{F}(\theta_0)^{-1}$. Actually, it is well known in time series analysis that the Fisher information matrix for the Gaussian process is

$$\mathscr{F}(\theta_0) = \frac{1}{4\pi} \int_{-\pi}^{\pi} f_{\theta_0}^{-2}(\lambda) \partial f_{\theta_0}(\lambda) \partial f_{\theta_0}(\lambda)^{\mathrm{T}} d\lambda,$$

which can be derived from the approximate maximum likelihood estimation. When the asymptotic variance of the estimator $\hat{\theta}_n$ attaining the Cramer–Rao lower bound, that is, the inverse matrix of Fisher information matrix $\mathscr{F}(\theta)^{-1}$, the estimator $\hat{\theta}_n$ is called asymptotically Gaussian efficient. We compare the asymptotic variances of the estimator $\hat{\theta}_n$ based on the disparity (2.24) in the following. In addition, an analytic lower bound for the estimator $\hat{\theta}_n$ is found in the following theorem.

Theorem 2.14 *Suppose Assumptions 2.2, 2.7, 2.8 and 2.12 hold. We obtain the following inequality in the matrix sense:*

$$H(\theta_0)^{-1}\tilde{V}(\theta_0)H(\theta_0)^{-1} \geq \mathscr{F}(\theta_0)^{-1}. \qquad (2.34)$$

The equality holds if $\alpha = -1$ or the spectral density $f(\omega)$ is a constant function.

Proof Define

$$A(\omega) = A_1(\theta_0)B_1(\theta_0) - A_2(\theta_0)B_2(\theta_0),$$
$$B(\omega) = \partial f_\theta(\omega)\Big|_{\theta=\theta_0},$$
$$g(\omega) = f_\theta(\omega)^2\Big|_{\theta=\theta_0}.$$

Then the inequality of (2.34) in Theorem 2.14 holds from Lemma 2.2. According to (2.19), the equality holds when

$$\int_{-\pi}^{\pi} f_\theta^{\alpha+1}(\lambda)d\lambda \, f_\theta^{\alpha+1}(\omega)\partial f_\theta(\omega)$$

$$- \int_{-\pi}^{\pi} f_\theta^\alpha(\lambda)\partial f_\theta(\lambda)d\lambda \, f_\theta^{\alpha+2}(\omega) - C\partial f_\theta(\omega)\Big|_{\theta=\theta_0} = 0 \quad (2.35)$$

with a generic constant C.

When $\alpha = -1$, then the left-hand side of (2.35) is

$$\int_{-\pi}^{\pi} f_\theta^{\alpha+1}(\lambda)d\lambda \, f_\theta^{\alpha+1}(\omega)\partial f_\theta(\omega)$$

$$- \int_{-\pi}^{\pi} f_\theta^\alpha(\lambda)\partial f_\theta(\lambda)d\lambda \, f_\theta^{\alpha+2}(\omega) - C\partial f_\theta(\omega)\Big|_{\theta=\theta_0}$$

$$= 2\pi \partial f_\theta(\omega) - \int_{-\pi}^{\pi} f_\theta^{-1}(\lambda)\partial f_\theta(\lambda)d\lambda \, f_\theta(\omega) - C\partial f_\theta(\omega)\Big|_{\theta=\theta_0}.$$

From (2.26), we have

$$\int_{-\pi}^{\pi} f_\theta^{-1}(\lambda)\partial f_\theta(\lambda)d\lambda\Big|_{\theta=\theta_0} = 0.$$

If we choose $C = 2\pi$, then the equality holds.

If $\alpha \neq -1$, then the equality in (2.34) does not hold in general. It is easy to see that (2.19) holds if the spectral density $f(\omega)$ does not depend on ω. □

In the following, let us illustrate Theorem 2.14 with two examples. Especially, we compare the asymptotic variance of the estimator $\hat{\theta}_n$ based on the disparity (2.24) when $\alpha = -2$ with that when $\alpha = -1$.

Example 2.4 Let $\{X(t)\}$ be generated by the AR(1) model as follows:

$$X(t) = \theta X(t-1) + \varepsilon(t), \quad |\theta| < 1, \quad \varepsilon(t) \sim \text{i.i.d. } \mathcal{N}(0, \sigma^2).$$

The spectral density $f_\theta(\omega)$ of $\{X(t)\}$ is expressed as

$$f_\theta(\omega) = \frac{\sigma^2}{2\pi} \frac{1}{|1 - \theta e^{i\omega}|^2}.$$

From Theorem 2.9, let $\alpha = -2$ and we obtain

$$H(\theta) = 2 \cdot (2\pi)^4 (1 - \theta^2), \quad V(\theta) = 4 \cdot (2\pi)^8 (1 - \theta^2)^2.$$

Thus, the asymptotic variance when $\alpha = -2$ is

$$H(\theta)^{-1} V(\theta) H(\theta)^{-1} = 1. \tag{2.36}$$

On the other hand, by (2.28), it holds that

$$\mathscr{F}(\theta) = \frac{1}{1 - \theta^2}. \tag{2.37}$$

Then comparing Eq. (2.36) with (2.37), we have

$$1 = H(\theta)^{-1} V(\theta) H(\theta)^{-1} \geq \mathscr{F}(\theta)^{-1} = 1 - \theta^2. \tag{2.38}$$

From (2.38), we can see that $\hat{\theta}_n$ is not asymptotically efficient except for $\theta = 0$.

Example 2.5 Let $\{X(t)\}$ be generated by the MA(1) model as follows:

$$X(t) = \varepsilon(t) + \theta \varepsilon(t-1), \quad |\theta| < 1, \quad \varepsilon_t \sim \text{i.i.d. } \mathcal{N}(0, \sigma^2).$$

The spectral density $f_\theta(\omega)$ of $\{X_t\}$ is

$$f_\theta(\omega) = \frac{\sigma^2}{2\pi} |1 + \theta e^{i\omega}|^2.$$

From Theorem 2.9, let $\alpha = -2$ and we obtain

$$H(\theta)^{-1} V(\theta) H(\theta)^{-1} = 1 - \theta^4 = (1 - \theta^2)(1 + \theta^2). \tag{2.39}$$

On the other hand, by (2.28), it holds that

$$\mathscr{F}(\theta) = \frac{1}{1 - \theta^2}. \tag{2.40}$$

Then comparing Eq. (2.39) with (2.40), we have

$$(1 - \theta^2)(1 + \theta^2) = H(\theta)^{-1}V(\theta)H(\theta)^{-1} \geq \mathscr{F}(\theta)^{-1} = 1 - \theta^2.$$

Therefore $\hat{\theta}_n$ is not asymptotically efficient except for $\theta = 0$.

2.4.3   Robustness Against Randomly Missing Observations

Not only robust against the fourth-order cumulant, the estimation by the disparity
(2.24) is also robust against randomly missing observations. This property is illustrated with some numerical simulations in Sect. 2.5. For conceptual understanding,
let us consider the amplitude modulated series, which is considered in Bloomfield
(1970). Let $\{Y(t)\}$ be an amplitude modulated series, that is,

$$Y(t) = X(t)Z(t),$$

where

$$Z(t) = \begin{cases} 1, & Y(t) \text{ is observed} \\ 0, & \text{otherwise.} \end{cases}$$

If we define $P(Z(t) = 1) = q$ and $P(Z(t) = 0) = 1 - q$, then the spectral density
f_Y for the series $\{Y(t)\}$ is represented by

$$f_Y(\omega) = q^2 f_X(\omega) + q \int_{-\pi}^{\pi} a(\omega - \alpha) f_X(\alpha) d\alpha, \tag{2.41}$$

where $a(\omega) = (2\pi)^{-1} \sum_r a_r e^{ir\omega}$ with $a_r = q^{-1} \text{Cov}(Z(t), Z(t+r))$. The spectral
density f_Y from Eq. (2.41) can be considered as the original spectral density f_X
heavily contaminated by a missing spectral density $\int_{-\pi}^{\pi} a(\omega - \alpha) f_X(\alpha) d\alpha$. (Basu
et al. 1998) used a similar minimum contrast estimator as a divergence for parameter
estimation for probability density function. They discussed in detail the trade-off
between robustness and efficiency in their paper. As pointed out in Fujisawa and
Eguchi (2008), the disparity has the robustness to outliers and contamination under
the heavy contaminated models.

2.5 Numerical Studies

In this section, we perform two numerical simulation examples for minimum contrast estimators based on our new disparity (2.17) for parameter estimation. The first simulation in Sect. 2.5.1 is to see the relative efficiency and the second simulation in Sect. 2.5.2 is to see the robustness of the estimator based on the disparity.

2.5.1 Relative Efficiency

In the first simulation, we investigate the empirical relative efficiency between the different choices of α when the true model is specified. Let us consider the AR(1) process

$$X(t) - bX(t-1) = \varepsilon(t), \tag{2.42}$$

where the innovation process $\{\varepsilon(t) : t = 1, 2, \ldots\}$ is assumed to be independent and identically distributed random variables. The distributions of innovations are assumed to be Gaussian, Laplace, and standard Cauchy. Laplace and Cauchy distributions are used as examples of a distribution with finite fourth-order cumulant and with infinite variance, respectively. All distributions are symmetric around 0. Gaussian distribution and Laplace distribution are set to have unit variance. We are interested in estimating the coefficient b in the model (2.42) by the disparity (2.24) with different values of α. The spectral density with parameter θ is assumed to be

$$f_\theta(\omega) = \frac{1}{2\pi}|1 - \theta e^{-i\omega}|^{-2}. \tag{2.43}$$

We generate 100 observations from the model (2.42) with coefficients $b = 0$ and 0.9. The estimation for θ by the disparity (2.24) is repeated via 100 Monte Carlo simulations. Let $\hat{\theta}_\alpha^{(i)}$ be the estimate by the exotic disparity with α in ith simulation. We define the empirical relative efficiency (ERE) by

$$\text{ERE} = \frac{\sum_{i=1}^{100}(\hat{\theta}_{-1}^{(i)} - b)^2}{\sum_{i=1}^{100}(\hat{\theta}_\alpha^{(i)} - b)^2}.$$

We use ERE as the benchmark since the Fisher information for the Gaussian process is achieved when $\alpha = -1$. The larger the index of ERE is, the better the performance of the estimation is. EREs for $b = 0$ and 0.9 are reported in Tables 2.1 and 2.2, respectively.

For the case $b = 0$, it is not difficult to see that the disparities achieve more than 74% in relative efficiency up to $\alpha = -3$. Without innovations distributed as Cauchy, the disparities with $\alpha = -4$ achieve 44% in relative efficiency.

Table 2.1 The empirical relative efficiency of the AR(1) model when $b = 0$

	$\alpha = -1$	$\alpha = -2$	$\alpha = -3$	$\alpha = -4$	$\alpha = -5$	$\alpha = -6$	$\alpha = -7$	$\alpha = -8$
Gaussian	1.000	0.955	0.853	0.447	0.091	0.061	0.043	0.038
Laplace	1.000	0.910	0.749	0.441	0.099	0.050	0.042	0.037
Cauchy	1.000	0.990	0.918	0.074	0.057	0.029	0.020	0.018

Table 2.2 The empirical relative efficiency of the AR(1) model when $b = 0.9$

	$\alpha = -1$	$\alpha = -2$	$\alpha = -3$	$\alpha = -4$	$\alpha = -5$	$\alpha = -6$	$\alpha = -7$	$\alpha = -8$
Gaussian	1.000	0.604	0.203	0.097	0.067	0.051	0.042	0.037
Laplace	1.000	0.578	0.162	0.075	0.047	0.035	0.029	0.025
Cauchy	1.000	0.847	0.304	0.118	0.073	0.057	0.048	0.041

On the other hand, it seems that the relative efficiencies of the disparities in Cauchy case work better than the other two finite variance cases when $b = 0.9$. However, the overall relative efficiencies are not as good as the case $b = 0$. Theoretically, the estimation for larger $|b| < 1$ has smaller asymptotic variance so the approaches in frequency domain are robust against unit root processes. Thus the Whittle estimator ($\alpha = -1$) has much better performance in this case. Nevertheless, the disparities achieve more than 57% in relative efficiency up to $\alpha = -2$.

2.5.2 Robustness of the Disparity

In the second simulation, we compare the robustness of the disparity (2.24) when α takes different values. Let us consider the parameter estimation by the disparity when the observations are randomly missing. To precisely formalize the problem, let $t \in \{1, \ldots, n\}$ and we are interested in parameter estimation by partial observations $\{X(t_i) : t_i \in \{1, \ldots, n\}\}$ of the following models.

(i) The same model as (2.42), i.e.,

$$X(t) - bX(t - 1) = \varepsilon(t).$$

(ii) The following MA(3) model:

$$X(t) = \varepsilon(t) + \frac{5}{4}\varepsilon(t - 1) + \frac{4}{3}\varepsilon(t - 2) + \frac{5}{6}\varepsilon(t - 3). \qquad (2.44)$$

If we fit (2.43) to the model (2.44), the disparity (2.17) is minimized by $\theta = 0.7$. This procedure of the parameter estimation can be regarded as a model misspecification case. In these two models, we suppose the innovation process $\{\varepsilon(t) : t =$

$1, \ldots, n\}$ is independent and identically distributed as the standard normal distribution and the standard Cauchy distribution as above.

First, we suppose that there are regularly missing points in our observations. Let $t_1 = 1$ and the observation period $d = t_{i+1} - t_i$ for all $1 \leq i \leq \lceil n/d \rceil$. We evaluate the robustness by the ratio of mean squared error of all estimation procedure with $\alpha = -1, -2, -3, -4$ and $d = 2, 3, 4$, where the ratio of mean squared error is defined by

$$\text{RoMSE} = \frac{\sum_{i=1}^{100} (\hat{\theta}_{-1}^{(i)} - b)^2}{\sum_{i=1}^{100} (\hat{\theta}_{\alpha}^{(i)} - b)^2}.$$

The simulation results for the model (2.42) are reported in Tables 2.3, 2.4, 2.5 and 2.6. The estimates are also given in the parentheses for each case.

Tables 2.3 and 2.4 show the ratio of mean squared error of the AR(1) model when $b = 0.9$ with the Cauchy and Gaussian innovations. It can be seen that the larger d is, the less efficient the Whittle estimator is, compared with other choices of α.

Tables 2.5 and 2.6 show the ratio of mean squared error of the AR(1) model when $b = 0.1$ with the Cauchy and Gaussian innovations. The similar aspect as in the case of $b = 0.9$ can be found when the AR(1) model has infinite variance innovations. On the other hand, when the innovations become Gaussian, the simulation results appear better for the Whittle estimates. One reason for this feature can be explained by the fact that the observations of AR(1) process for a long period are almost independent.

Next, we generated 128 observations from the model (2.42) and (2.44), respectively. We randomly chose four sets of time points $T_i \subset \{1, 2, \ldots, 128\}$ as the observed time points as follows. The length of T_i is 32 for $i = 1, 2, 3, 4$.

Table 2.3 The ratio of mean squared error and estimate (in parentheses) of the AR(1) model with the Cauchy innovations when $b = 0.9$

Cauchy	$\alpha = -1$	$\alpha = -2$	$\alpha = -3$	$\alpha = -4$
$d = 2$	1.000 (0.745)	1.248 (0.782)	0.757 (0.806)	0.457 (0.772)
$d = 3$	1.000 (0.648)	1.345 (0.717)	1.303 (0.786)	0.992 (0.778)
$d = 4$	1.000 (0.549)	1.195 (0.609)	1.257 (0.677)	1.143 (0.700)

Table 2.4 The ratio of mean squared error and estimate (in parentheses) of the AR(1) model with the Gaussian innovations when $b = 0.9$

Gaussian	$\alpha = -1$	$\alpha = -2$	$\alpha = -3$	$\alpha = -4$
$d = 2$	1.000 (0.727)	1.138 (0.784)	0.829 (0.796)	0.548 (0.772)
$d = 3$	1.000 (0.609)	1.194 (0.681)	1.088 (0.730)	0.835 (0.700)
$d = 4$	1.000 (0.511)	1.186 (0.585)	1.278 (0.651)	1.121 (0.642)

Table 2.5 The ratio of mean squared error and estimate (in parentheses) of the AR(1) model with the Cauchy innovations when $b = 0.1$

Cauchy	$\alpha = -1$	$\alpha = -2$	$\alpha = -3$	$\alpha = -4$
$d = 2$	1.000 (0.455)	4.140 (0.721)	3.045 (0.992)	3.145 (0.998)
$d = 3$	1.000 (0.148)	1.049 (0.195)	1.086 (0.288)	1.118 (0.374)
$d = 4$	1.000 (−0.005)	1.004 (−0.001)	1.017 (0.041)	0.858 (0.024)

Table 2.6 The ratio of mean squared error and estimate (in parentheses) of the AR(1) model with the Gaussian innovations when $b = 0.1$

Gaussian	$\alpha = -1$	$\alpha = -2$	$\alpha = -3$	$\alpha = -4$
$d = 2$	1.000 (0.434)	2.783 (0.674)	4.449 (0.969)	4.377 (0.981)
$d = 3$	1.000 (0.122)	1.029 (0.138)	1.064 (0.226)	1.316 (0.366)
$d = 4$	1.000 (−0.038)	0.958 (−0.048)	0.749 (−0.089)	0.678 (−0.071)

Table 2.7 The ratio of mean squared error of the AR(1) model with the Cauchy innovations when $b = 0.9$

Cauchy	$\alpha = -1$	$\alpha = -2$	$\alpha = -3$	$\alpha = -4$
T_1	1.000	1.192	1.149	1.007
T_2	1.000	1.332	1.294	1.053
T_3	1.000	1.182	1.158	0.948
T_4	1.000	1.288	1.307	1.073

$$T_1 = \{1, 7, 9, 17, 19, 23, 26, 30, 34, 39, 44, 50, 54, 58, 59, 61, 66, 67,$$
$$74, 75, 79, 80, 81, 87, 101, 102, 104, 112, 118, 121, 125, 128\},$$
$$T_2 = \{1, 2, 4, 6, 9, 13, 21, 22, 31, 36, 37, 38, 39, 42, 49, 50, 56, 71,$$
$$76, 77, 82, 85, 93, 96, 101, 110, 112, 113, 115, 117, 126, 127\},$$
$$T_3 = \{1, 4, 5, 12, 14, 15, 18, 20, 23, 26, 27, 28, 33, 39, 41, 55, 56, 74,$$
$$78, 83, 84, 85, 88, 100, 104, 106, 107, 108, 109, 114, 115, 120\},$$
$$T_4 = \{2, 4, 12, 14, 34, 38, 39, 41, 42, 43, 44, 49, 54, 55, 56, 59, 60, 63,$$
$$65, 70, 72, 78, 81, 94, 98, 100, 103, 107, 110, 119, 123, 126\}.$$

We take observations on the sets T_i ($i = 1, \ldots, 4$) for the parameter estimation by the disparity (2.24) The ratio of mean squared errors via 100 simulations are given in Tables 2.7, 2.8, 2.9 and 2.10.

From Tables 2.7 and 2.8, one can see that the mean squared errors of $\alpha = -2$ and $\alpha = -3$ for $T_1 - T_4$ are smaller than that of $\alpha = -1$. Even $\alpha = -4$, the mean squared error is larger than $\alpha = -1$ only for T_3. In the misspecification case of model

Table 2.8 The ratio of mean squared error of the AR(1) model with the Gaussian innovations when $b = 0.9$

Gaussian	$\alpha = -1$	$\alpha = -2$	$\alpha = -3$	$\alpha = -4$
T_1	1.000	1.287	1.490	1.435
T_2	1.000	1.181	1.220	1.199
T_3	1.000	1.148	1.247	1.078
T_4	1.000	1.230	1.395	1.125

Table 2.9 The ratio of mean squared error of the MA(3) model (2.44) with the Cauchy innovations

Cauchy	$\alpha = -1$	$\alpha = -2$	$\alpha = -3$	$\alpha = -4$
T_1	1.000	1.053	1.093	0.959
T_2	1.000	1.107	1.107	1.230
T_3	1.000	1.129	1.266	1.260
T_4	1.000	1.068	1.078	0.961

Table 2.10 The ratio of mean squared error of the MA(3) model (2.44) with the Gaussian innovations

Gaussian	$\alpha = -1$	$\alpha = -2$	$\alpha = -3$	$\alpha = -4$
T_1	1.000	1.061	1.140	1.181
T_2	1.000	1.099	1.294	2.085
T_3	1.000	1.136	1.605	1.533
T_4	1.000	1.052	1.144	1.291

(2.44), $\alpha = -2$ and $\alpha = -3$ also perform better in the sense of mean squared error than $\alpha = -1$. In conclusion, an appropriate choice of α leads to robust parameter estimation to randomly missing observations.

Chapter 3
Quantile Method for Time Series

Abstract In this chapter, we introduce a nonparametric method to statistically investigate stationary time series. We have seen that there exists a spectral distribution function for any second-order stationary process. We define quantiles of the spectral distribution function in the frequency domain and consider the quantile method for parameter estimation of stationary time series. The estimation method for quantiles is generally formulated by minimizing a check function. The quantile estimator is shown to be asymptotically normal. We also consider the hypothesis testing problem for quantiles in the frequency domain and propose a test statistic associated with our quantile estimator, which asymptotically converges to the standard normal under the null hypothesis. The finite sample performance of the quantile estimator is shown in our numerical studies.

3.1 Introduction

Nowadays, the quantile-based estimation becomes a notable method in statistics. Not only statistical inference for the quantile of the cumulative distribution function is considered, the quantile regression, a method taking place of the ordinary regression, is also broadly used for statistical inference. (See Koenker 2005.) In the area of time series analysis, however, the quantile-based inference is still undeveloped yet. A fascinating approach in the frequency domain, called "quantile periodogram" is proposed and studied in Li (2008, 2012). More general methods associated with copulas, quantiles, and ranks are developed in Dette et al. (2015), Kley et al. (2016) and Birr et al. (2017). Especially, a quantile-based method for time series analysis is discussed in Liu (2017b).

As there exists a well-behaved spectral distribution function for a second-order stationary process, we introduce the quantiles of the spectral distribution and develop a statistical inference theory for it. For a second-order stationary process with continuous spectral distribution, we consider a quantile estimator defined by the minimizer

© The Author(s), under exclusive licence to Springer Nature Singapore Pte Ltd. 2018   59
Y. Liu et al., *Empirical Likelihood and Quantile Methods for Time Series*,
JSS Research Series in Statistics, https://doi.org/10.1007/978-981-10-0152-9_3

of check function based on the periodogram for estimation. The consistency and the asymptotic normality of the estimator are shown under a natural assumption that the spectral density is sufficiently smooth. We also propose a quantile test in the frequency domain to test the dependence structure of second-order stationary processes, since the spectral distribution function is uniquely determined by the autocovariance function of the process. When the spectral distributions of processes considered are quite different, the testing procedure works well.

In the context of time series analysis, (Whittle 1952b) mentioned that "the search for periodicities" constituted the whole of time series theory. He proposed an estimation method based on a nonlinear model driven by a simple harmonic component. After the work, to estimate the frequency has been a remarkable statistical analysis. A sequential literature by Whittle (1952b), Walker (1971), Hannan (1973), Rice and Rosenblatt (1988), Quinn and Thomson (1991), and Quinn and Hannan (2001) investigated the method proposed by Whittle (1952b) and pointed out the misunderstandings in Whittle (1952b), respectively. The noise structure is also generalized from independent and identically distributed white noise to the second-order stationary process. On the other hand, statistical inference of the spectral density by parametric form is considered by Dzhaparidze (1986) and the reference therein. In our estimation procedure, we employ the check function to estimate quantiles, i.e., the frequencies of the spectral distribution function for second-order stationary processes. The method is different from all the methods mentioned above. With the order $O(\sqrt{n})$, the basic quantile estimators for continuous spectral distribution are shown to be asymptotically normal. It is applicable to implement the quantile tests in the frequency domain by our new consideration on the quantiles of the spectral distribution.

3.2 Preliminaries

In this section, we discuss properties of the quantiles of the spectral distribution. From Chap. 2, we have seen that the parametric spectral density f_θ for linear processes with finite variance (or power transfer function for linear processes with infinite variance) can be consistently estimated via minimum contrast estimation. The merits of estimating spectral quantiles can be summarized in the following points:

- A nonparametric method for statistical inference is more flexible than the parametric approach;
- A quantile-based estimation is simple in that the bandwidth choice is avoided, compared with a fully nonparametric method such as kernel estimation;
- Two spectral distributions (or power transfer functions) are possible to be discriminated by their own quantiles without any loss in the rate of convergence;
- The method is robust to outliners, especially in the sense that there are missing observations of the process.

The quantile λ_p in (1.10) can be equivalently defined as a minimizer of some certain objective function as follows. Suppose $\{X(t) : t \in \mathbb{Z}\}$ is a zero-mean second-order stationary process with finite autocovariance function $R_X(h) = \text{Cov}(X(t + h), X(t))$, for $h \in \mathbb{Z}$. From Theorem 1.1 (Herglotz's theorem), the process $\{X(t)\}$ has a spectral distribution $F_X(\omega)$.

Let Λ be $\Lambda = [-\pi, \pi]$. In the following, we show that the pth quantile λ_p in (1.10) can be defined by the minimizer of the following objective function $S(\theta)$, i.e.,

$$S(\theta) = \int_{-\pi}^{\pi} \rho_p(\omega - \theta) F_X(d\omega), \tag{3.1}$$

where $\rho_\tau(u)$, called "the check function" (e.g. Koenker 2005), is defined as

$$\rho_\tau(u) = u(\tau - \mathbb{1}(u < 0)),$$

where $\mathbb{1}(\cdot)$ is the indicator function.

Theorem 3.1 *Suppose the process $\{X(t) : t \in \mathbb{Z}\}$ is a zero-mean second-order stationary process with spectral distribution function $F_X(\omega)$. Define $S(\theta)$ by (3.1). Then the pth quantile λ_p of the spectral distribution $F_X(\omega)$ is a minimizer of $S(\theta)$ on Λ. Furthermore, λ_p is unique and satisfies*

$$\lambda_p = \inf\{\omega \in \Lambda; \ S(\omega) = \min_{\theta \in \Lambda} S(\theta)\}. \tag{3.2}$$

Proof First, we confirm the existence of the minimizer of $S(\theta)$. The right derivative of $S(\theta)$ is

$$S'_+(\theta) \equiv \lim_{\varepsilon \to +0} \frac{S(\theta + \varepsilon) - S(\theta)}{\varepsilon} = F_X(\theta) - p\Sigma_X.$$

From (1.10), we have

$$S'_+(\theta) \begin{cases} < 0, & \text{for } \theta < \lambda_p, \\ \geq 0, & \text{for } \theta \geq \lambda_p. \end{cases}$$

Thus, the minimizer of $S(\theta)$ exists and $S(\lambda_p) = \min_{\theta \in \Lambda} S(\theta)$. The uniqueness of λ_p and the representation (3.2) follow from (1.10). $\qquad\square$

The representation (3.2) of the pth quantile λ_p of the spectral distribution function $F_X(\omega)$ is useful when we consider the estimation theory of λ_p. From the definition of the spectral distribution function, $F_X(\omega)$ is uniquely determined by the autocovariance function $R_X(h)$ ($h \in \mathbb{Z}$). Accordingly, the dependence structures of the second-order stationary processes $\{X(t)\}$ and $\{X'(t)\}$ can be discriminated by the pth quantile λ_p since $\lambda_p \neq \lambda'_p$ if $p \neq 0, 1/2, 1$, or $X(t) \neq cX'(t)$, $c \in \mathbb{R}$ for all $t \in \mathbb{Z}$.

3.3   Quantile Estimation

In this section, we consider a method of quantile estimation for spectral distributions. Suppose the process $\{X(t)\}$ has the spectral density $f_X(\omega)$, i.e.,

$$R_X(h) = \int_{-\pi}^{\pi} e^{-ih\omega} f_X(\omega) d\omega.$$

Denote by $\{X(t) : 1 \leq t \leq n\}$ the observation stretch of the process. The parameter space for the pth quantile λ_p is Λ. λ_p is in the interior of Λ. Let the periodogram $I_{n,X}(\omega)$ of the observation stretch be defined at the Fourier frequencies $\omega_s = 2\pi s/n$, $\omega_s \in \Lambda$, as

$$I_{n,X}(\omega_s) = \frac{1}{2\pi n} \left| \sum_{t=1}^{n} X(t) e^{it\omega_s} \right|^2.$$

For estimation problem, we define the sample objective function $S_n(\theta)$ as

$$S_n(\theta) = \frac{2\pi}{n} \sum_{\omega_s \in \Lambda, s \in \mathbb{Z}} \rho_p(\omega_s - \theta) I_{n,X}(\omega_s), \tag{3.3}$$

and the quantile estimator $\hat{\lambda}_p$ as

$$\hat{\lambda}_p \equiv \hat{\lambda}_{p,n} = \arg\min_{\theta \in \Lambda} S_n(\theta). \tag{3.4}$$

Remark 3.1 As in (3.3), we define the minimization problem by the sum over all Fourier frequencies $\omega_s \in \Lambda$. That is, for fixed n, the summation is over all integers s satisfying $-\pi \leq (2\pi s)/n \leq \pi$. To simplify the notation, we use $\sum_{\omega_s \in \Lambda}$ instead of $\sum_{\omega_s \in \Lambda, s \in \mathbb{Z}}$, where there should be no confusion. Henceforth, the sum $\sum_{\omega_s \in \Lambda}$ is always defined over all Fourier frequencies ω_s.

Let us consider the asymptotic properties of the estimator $\hat{\lambda}_p$ defined in (3.4) for stationary process $\{X(t) : t \in \mathbb{Z}\}$ under the following assumptions.

Assumption 3.2 $\{X(t)\}$ is a zero-mean, strictly stationary real-valued process, all of whose moments exist with

$$\sum_{u_1, \ldots, u_{k-1} = -\infty}^{\infty} |\mathrm{cum}_X(u_1, \ldots, u_{k-1})| < \infty, \quad \text{for } k = 2, 3, \ldots,$$

where $\mathrm{cum}_X(u_1, \ldots, u_{k-1})$ denotes the joint cumulant of $\big(X(t), X(t + u_1), \ldots, X(t + u_{k-1})\big)$.

Under Assumption 3.2, the fourth-order spectral density is defined as

$$Q_X(\omega_1, \omega_2, \omega_3) = \frac{1}{(2\pi)^3} \sum_{t_1,t_2,t_3=-\infty}^{\infty} \exp\{-i(\omega_1 t_1 + \omega_2 t_2 + \omega_3 t_3)\} \mathrm{cum}_X(t_1, t_2, t_3).$$

First, we show the consistency of the estimator $\hat{\lambda}_p$ under Assumption 3.2.

Theorem 3.3 *Suppose the process $\{X(t) : t \in \mathbb{Z}\}$ satisfies Assumption 3.2 and the pth quantile λ_p, defined by (1.10), satisfies $f_X(\lambda_p) > 0$. If $\hat{\lambda}_p$ is defined by (3.4), then we have*

$$\hat{\lambda}_p \overset{\mathscr{P}}{\longrightarrow} \lambda_p.$$

Proof We first show the sample objective function $S_n(\theta)$ is convex in θ. That is, for $0 \le t \le 1$,

$$S_n(t\theta_1 + (1-t)\theta_2) \le t S_n(\theta_1) + (1-t)S_n(\theta_2). \tag{3.5}$$

The left-hand side of (3.5) has the following expression:

$$S_n(t\theta_1 + (1-t)\theta_2)$$

$$= \frac{2\pi}{n} \sum_{\omega_s \in \Lambda} \mathbb{1}(\omega_s < \theta_1, \omega_s < \theta_2)\big(\omega_s - (t\theta_1 + (1-t)\theta_2)\big)$$

$$\times \big(p - \mathbb{1}(\omega_s < t\theta_1 + (1-t)\theta_2)\big) I_{n,X}(\omega_s)$$

$$+ \frac{2\pi}{n} \sum_{\omega_s \in \Lambda} \Big(1 - \mathbb{1}(\omega_s < \theta_1, \omega_s < \theta_2)\Big)\big(\omega_s - (t\theta_1 + (1-t)\theta_2)\big)$$

$$\times \big(p - \mathbb{1}(\omega_s < t\theta_1 + (1-t)\theta_2)\big) I_{n,X}(\omega_s)$$

$$= S_n^1(t) + S_n^2(t), \quad \text{(say)}.$$

The inequality (3.5) holds if we show $S_n^1(t)$ and $S_n^2(t)$ have the corresponding expression, respectively.
We first consider $S_n^1(t)$.

$$S_n^1(t) = \frac{2\pi}{n} \sum_{\omega_s \in \Lambda} \mathbb{1}(\omega_s < \theta_1, \omega_s < \theta_2)\big(t(\omega_s - \theta_1) + (1-t)(\omega_s - \theta_2)\big)$$

$$\times \big(p - \mathbb{1}(0 < t(\theta_1 - \omega_s) + (1-t)(\theta_2 - \omega_s))\big) I_{n,X}(\omega_s)$$

$$= \frac{2\pi}{n} \sum_{\omega_s \in \Lambda} \mathbb{1}(\omega_s < \theta_1, \omega_s < \theta_2)$$

$$\times \big(t(\omega_s - \theta_1) + (1-t)(\omega_s - \theta_2)\big)(p-1) I_{n,X}(\omega_s)$$

$$= \frac{2\pi}{n} \sum_{\omega_s \in \Lambda} \mathbb{1}(\omega_s < \theta_1, \omega_s < \theta_2)\big(t(\omega_s - \theta_1)(p - \mathbb{1}(\omega_s < \theta_1))$$

$$+ (1-t)(\omega_s - \theta_2)\big(p - \mathbb{1}(\omega_s < \theta_2)\big)\big) I_{n,X}(\omega_s)$$

$$
= t \cdot \frac{2\pi}{n} \sum_{\omega_s \in \Lambda} \mathbb{1}(\omega_s < \theta_1, \omega_s < \theta_2) \rho_p(\omega_s - \theta_1) I_{n,X}(\omega_s)
$$

$$
+ (1 - t) \cdot \frac{2\pi}{n} \sum_{\omega_s \in \Lambda} \mathbb{1}(\omega_s < \theta_1, \omega_s < \theta_2) \rho_p(\omega_s - \theta_2) I_{n,X}(\omega_s)
$$

$$
= \tilde{S}_n^1(t), \quad \text{(say)}.
$$

Next, we consider $S_n^2(t)$. Note that for any $a, b \in \mathbb{R}$, $\mathbb{1}(0 < a + b) \le \mathbb{1}(0 < a) + \mathbb{1}(0 < b)$. For $0 \le t \le 1$,

$$
S_n^2(t) = \frac{2\pi}{n} \sum_{\omega_s \in \Lambda} \left(1 - \mathbb{1}(\omega_s < \theta_1, \omega_s < \theta_2)\right)\left(\omega_s - (t\theta_1 + (1 - t)\theta_2)\right)
$$

$$
\times \left(p - \mathbb{1}(\omega_s < t\theta_1 + (1 - t)\theta_2)\right) I_{n,X}(\omega_s)
$$

$$
= \frac{2\pi}{n} \sum_{\omega_s \in \Lambda} \left(1 - \mathbb{1}(\omega_s < \theta_1, \omega_s < \theta_2)\right)
$$

$$
\times \left\{\left(\omega_s - (t\theta_1 + (1 - t)\theta_2)\right) p I_{n,X}(\omega_s) - \left(\omega_s - (t\theta_1 + (1 - t)\theta_2)\right)\right.
$$

$$
\left. \times \left(\mathbb{1}(0 < t(\theta_1 - \omega_s) + (1 - t)(\theta_2 - \omega_s))\right) I_{n,X}(\omega_s)\right\}
$$

$$
\le \frac{2\pi}{n} \sum_{\omega_s \in \Lambda} \left(1 - \mathbb{1}(\omega_s < \theta_1, \omega_s < \theta_2)\right)
$$

$$
\times \left\{\left(t(\omega_s - \theta_1) + (1 - t)(\omega_s - \theta_2)\right) p I_{n,X}(\omega_s)\right.
$$

$$
- \left(t(\omega_s - \theta_1) + (1 - t)(\omega_s - \theta_2)\right)
$$

$$
\left. \times \left(\mathbb{1}(0 < t(\theta_1 - \omega_s)) + \mathbb{1}(0 < (1 - t)(\theta_2 - \omega_s))\right) I_{n,X}(\omega_s)\right\}
$$

$$
\le t \cdot \frac{2\pi}{n} \sum_{\omega_s \in \Lambda} \left(1 - \mathbb{1}(\omega_s < \theta_1, \omega_s < \theta_2)\right)(\omega_s - \theta_1)
$$

$$
\times \left(p - \left(\mathbb{1}(0 < \theta_1 - \omega_s)\right)\right) I_{n,X}(\omega_s)
$$

$$
+ (1 - t) \cdot \frac{2\pi}{n} \sum_{\omega_s \in \Lambda} \left(1 - \mathbb{1}(\omega_s < \theta_1, \omega_s < \theta_2)\right)(\omega_s - \theta_2)
$$

$$
\times \left(p - \left(\mathbb{1}(0 < \theta_1 - \omega_s)\right)\right) I_{n,X}(\omega_s)
$$

$$
- \frac{2\pi}{n} \sum_{\omega_s \in \Lambda} \left(1 - \mathbb{1}(\omega_s < \theta_1, \omega_s < \theta_2)\right) t(\omega_s - \theta_1) \mathbb{1}(0 < \theta_2 - \omega_s) I_{n,X}(\omega_s)
$$

$$
- \frac{2\pi}{n} \sum_{\omega_s \in \Lambda} \left(1 - \mathbb{1}(\omega_s < \theta_1, \omega_s < \theta_2)\right)
$$

$$
\times (1 - t)(\omega_s - \theta_2) \mathbb{1}(0 < \theta_1 - \omega_s) I_{n,X}(\omega_s)
$$

$$\leq t \cdot \frac{2\pi}{n} \sum_{\omega_s \in \Lambda} \left(1 - \mathbb{1}(\omega_s < \theta_1, \omega_s < \theta_2)\right) \rho_p(\omega_s - \theta_1) I_{n,X}(\omega_s)$$

$$+ (1-t) \cdot \frac{2\pi}{n} \sum_{\omega_s \in \Lambda} \left(1 - \mathbb{1}(\omega_s < \theta_1, \omega_s < \theta_2)\right) \rho_p(\omega_s - \theta_2) I_{n,X}(\omega_s) \quad (3.6)$$

$$= \tilde{S}_n^2(t), \quad \text{(say)}.$$

The last inequality ($\leq$) of (3.6) holds since
(i) if $(\theta_2 - \omega_s)(\theta_1 - \omega_s) \leq 0$, then it holds that

$$-\frac{2\pi}{n} \sum_{\omega_s \in \Lambda} \left(1 - \mathbb{1}(\omega_s < \theta_1, \omega_s < \theta_2)\right) t\left(\omega_s - \theta_1\right) \mathbb{1}\left(0 < \theta_2 - \omega_s\right) I_{n,X}(\omega_s)$$

$$-\frac{2\pi}{n} \sum_{\omega_s \in \Lambda} \left(1 - \mathbb{1}(\omega_s < \theta_1, \omega_s < \theta_2)\right)$$

$$\times (1-t)\left(\omega_s - \theta_2\right) \mathbb{1}\left(0 < \theta_1 - \omega_s\right) I_{n,X}(\omega_s) \leq 0, \quad \text{a.s.};$$

(ii) if $(\theta_2 - \omega_s)(\theta_1 - \omega_s) > 0$, then it holds

$$-\frac{2\pi}{n} \sum_{\omega_s \in \Lambda} \left(1 - \mathbb{1}(\omega_s < \theta_1, \omega_s < \theta_2)\right) t\left(\omega_s - \theta_1\right) \mathbb{1}\left(0 < \theta_2 - \omega_s\right) I_{n,X}(\omega_s)$$

$$-\frac{2\pi}{n} \sum_{\omega_s \in \Lambda} \left(1 - \mathbb{1}(\omega_s < \theta_1, \omega_s < \theta_2)\right)$$

$$\times (1-t)\left(\omega_s - \theta_2\right) \mathbb{1}\left(0 < \theta_1 - \omega_s\right) I_{n,X}(\omega_s) = 0, \quad \text{a.s.}$$

In summary, combining $S_n^1(t) \leq \tilde{S}_n^1(t)$ and $S_n^2(t) \leq \tilde{S}_n^2(t)$, the inequality (3.5) holds. Hence, $\{S_n(\theta), \theta \in \Lambda\}$ is a sequence of random convex functions in θ.

Now, let us consider the pointwise limit of $S_n(\theta)$. Actually, for each $\theta \in \Lambda$,

$$|S_n(\theta) - S(\theta)|$$

$$\leq \left| \frac{2\pi}{n} \sum_{\omega_s \in \Lambda} \rho_p(\omega_s - \theta) I_{n,X}(\omega_s) - E\left(\frac{2\pi}{n} \sum_{\omega_s \in \Lambda} \rho_p(\omega_s - \theta) I_{n,X}(\omega_s)\right) \right|$$

$$+ \left| E\left(\frac{2\pi}{n} \sum_{\omega_s \in \Lambda} \rho_p(\omega_s - \theta) I_{n,X}(\omega_s)\right) - \int_{-\pi}^{\pi} \rho_p(\omega - \theta) f_X(\omega) d\omega \right|. \quad (3.7)$$

From Theorem 7.6.1 in Brillinger (2001), it holds under Assumption 3.2 that

$$\text{Var}\left(\frac{2\pi}{n} \sum_{\omega_s \in \Lambda} \rho_p(\omega_s - \theta) I_{n,X}(\omega_s)\right) = O(n^{-1}).$$

The first term of the right-hand side in (3.7) converges to 0 in probability, which can be shown by Chebyshev's inequality.

Again, from Theorem 7.6.1 in Brillinger (2001), it holds under Assumption 3.2 that

$$E\left(\frac{2\pi}{n}\sum_{\omega_s\in\Lambda}\rho_p(\omega_s-\theta)I_{n,X}(\omega_s)\right) = \int_{-\pi}^{\pi}\rho_p(\omega-\theta)f_X(\omega)\,d\omega + o(1),$$

since the check function $\rho_p(\omega-\theta)$ is of bounded variation. Hence, the second term of the right-hand side in (3.7) converges to 0. Therefore, from (3.7), we see that for each $\theta\in\Lambda$,

$$S_n(\theta)\xrightarrow{\mathscr{P}}S(\theta).$$

By the Convexity Lemma in Pollard (1991), it holds that

$$\sup_{\theta\in K}|S_n(\theta)-S(\theta)|\xrightarrow{\mathscr{P}}0,\tag{3.8}$$

for any compact subset $K\subset\Lambda$.

Now, let us consider λ_p, which is the only minimizer of $S(\theta)$ since $f_X(\lambda_p)>0$. Let m be the minimum of $S(\theta)$, and $B(\lambda_p)$ be any open neighborhood of λ_p. From the uniqueness of minimizer of $S(\theta)$, there exists an $\varepsilon>0$ such that $\inf_{\theta\in\Lambda\backslash B(\lambda_p)}|S(\theta)|>m+\varepsilon$. Thus, with probability tending to 1,

$$\inf_{\theta\in\Lambda\backslash B(\lambda_p)}S_n(\theta)\geq\inf_{\theta\in\Lambda\backslash B(\lambda_p)}S(\theta)-\sup_{\theta\in\Lambda\backslash B(\lambda_p)}|S(\theta)-S_n(\theta)|>m.\tag{3.9}$$

Note that the second inequality follows that by (3.8) and the second term can be chosen arbitrarily small, i.e.,

$$\sup_{\theta\in\Lambda\backslash B(\lambda_p)}|S(\theta)-S_n(\theta)|<\varepsilon.$$

On the other hand, with probability tending to 1,

$$S_n(\hat{\lambda}_p)\leq m+\varepsilon^*\tag{3.10}$$

for any $\varepsilon^*>0$, by the pointwise convergence of $S_n(\lambda_p)$ to $S(\lambda_p)(=m)$ in probability. By contradiction between (3.9) and (3.10), one sees that $\hat{\lambda}_p\in B(\lambda_p)$ with probability tending to 1, which completes the proof.   □

The consistency of the estimator $\hat{\lambda}_p$ is not difficult to expect. The result, however, requires the continuity of the spectral distribution function $F_X(\omega)$, a strong assumption, if we stand on the estimator (3.4). It is possible to modify the estimator (3.4) by smoothing to loose Assumption 3.2. The result will be elucidated in another work.

Next, we investigate the asymptotic distribution of the estimator $\hat{\lambda}_p$. We impose the following assumption on the process $\{X(t)\}$ instead of Assumption 3.2, which is stronger than Assumption 3.2.

Assumption 3.4 The process $\{X(t)\}$ is a zero-mean, strictly stationary real-valued process, all of whose moments exist with

$$\sum_{u_1,\dots,u_{k-1}=-\infty}^{\infty} \left(1 + \sum_{j=1}^{k-1} |u_j|\right) |\mathrm{cum}_X(u_1,\dots,u_{k-1})| < \infty, \quad \text{for } k = 2, 3, \dots.$$

The asymptotic distribution of $\hat{\lambda}_p$ is given as follows.

Theorem 3.5 *Suppose* $\{X(t) : t \in \mathbb{Z}\}$ *satisfies Assumption 3.4 and the pth quantile* λ_p, *defined by (1.10), satisfies* $f_X(\lambda_p) > 0$. *If* $\hat{\lambda}_p$ *is defined by (3.4), then we have*

$$\sqrt{n}\,(\hat{\lambda}_p - \lambda_p) \xrightarrow{\mathcal{L}} \mathcal{N}(0, \sigma_p^2),$$

where

$$\begin{aligned}
\sigma_p^2 = f_X(\lambda_p)^{-2}\bigg[& 4\pi p^2 \int_{-\pi}^{\pi} f_X(\omega)^2 d\omega \\
& + 2\pi(1 - 4p)\int_{-\pi}^{\lambda_p} f_X(\omega)^2 d\omega + 2\pi\int_{-\lambda_p}^{\lambda_p} f_X(\omega)^2 d\omega \\
& + 2\pi\bigg\{ \int_{-\pi}^{\lambda_p}\int_{-\pi}^{\lambda_p} Q_X(\omega_1, \omega_2, -\omega_2) d\omega_1 d\omega_2 \\
& + \int_{-\pi}^{\pi}\int_{-\pi}^{\pi} p^2 Q_X(\omega_1, \omega_2, -\omega_2) d\omega_1 d\omega_2 \\
& - 2p\int_{-\pi}^{\lambda_p}\int_{-\pi}^{\pi} Q_X(\omega_1, \omega_2, -\omega_2) d\omega_1 d\omega_2 \bigg\}\bigg].
\end{aligned}$$

To prove Theorem 3.5, we need to apply Corollary 2 in Knight (1998). (See also Geyer 1996.) To clarify the thread of the proof of Theorem 3.5, we list all the steps here:

(i) We decide the order of convergence in Lemma 3.1.
(ii) To find the distribution of $M(\delta)$ in Corollary 2 in Knight (1998), we derive the asymptotic joint distribution of quantities involved in $M_n(\delta)$.
(iii) Applying Slutsky's theorem to derive the asymptotic distribution $M(\delta)$.

In Lemma 3.1, we consider the asymptotic variance of

$$T_n(\lambda) \equiv n^{\beta} \cdot \left(\frac{2\pi}{n} \sum_{\lambda < \omega_s \leq \lambda + n^{-\beta}} I_{n,X}(\omega_s)\right). \tag{3.11}$$

The asymptotic variance can be classified as in the following lemma.

Lemma 3.1 *Suppose $\{X(t)\}$ satisfies Assumption 3.4. Let $T_n(\lambda)$ be defined as (3.11). Then the asymptotic variance of $T_n(\lambda)$ is given by*

$$\lim_{n \to \infty} \mathrm{Var}(T_n(\lambda)) = \begin{cases} 0, & \text{if } \beta < 1, \\ f_X(\lambda)^2, & \text{if } \beta = 1, \\ \infty, & \text{if } \beta > 1. \end{cases}$$

Proof Let $a_n := n^{\beta}$. Divide $T_n(\lambda)$ by

$$a_n \left(\frac{2\pi}{n} \sum_{-\pi < \omega_s \leq \lambda + a_n} I_{n,X}(\omega_s) \right) - a_n \left(\frac{2\pi}{n} \sum_{-\pi < \omega_s \leq \lambda} I_{n,X}(\omega_s) \right).$$

The variances of both two parts and their covariance in asymptotics are given by

$$\mathrm{Var} \left\{ a_n \left(\frac{2\pi}{n} \sum_{-\pi < \omega_s \leq \lambda + a_n} I_{n,X}(\omega_s) \right) \right\}$$

$$= \frac{a_n^2}{n} 2\pi \left(\int_{-\pi}^{\lambda + a_n^{-1}} f_X(\omega)^2 d\omega \right.$$

$$\left. + \int_{-\pi}^{\lambda + a_n^{-1}} \int_{-\pi}^{\lambda + a_n^{-1}} Q_X(\omega_1, \omega_2, -\omega_2) d\omega_1 d\omega_2 \right) + \text{lower order},$$

$$\mathrm{Var} \left\{ a_n \left(\frac{2\pi}{n} \sum_{-\pi < \omega_s \leq \lambda} I_{n,X}(\omega_s) \right) \right\}$$

$$= \frac{a_n^2}{n} 2\pi \left(\int_{-\pi}^{\lambda} f_X(\omega)^2 d\omega + \int_{-\pi}^{\lambda} \int_{-\pi}^{\lambda} Q_X(\omega_1, \omega_2, -\omega_2) d\omega_1 d\omega_2 \right) + \text{lower order},$$

and

$$\mathrm{Cov} \left(a_n \left(\frac{2\pi}{n} \sum_{-\pi < \omega_s \leq \lambda + a_n} I_{n,X}(\omega_s) \right), a_n \left(\frac{2\pi}{n} \sum_{-\pi < \omega_s \leq \lambda} I_{n,X}(\omega_s) \right) \right)$$

$$= \frac{a_n^2}{n} 2\pi \left(\int_{-\pi}^{\lambda} f_X(\omega)^2 d\omega + \int_{-\pi}^{\lambda} \int_{-\pi}^{\lambda + a_n^{-1}} Q_X(\omega_1, \omega_2, -\omega_2) d\omega_1 d\omega_2 \right) + \text{lower order}.$$

As a result, the variance of $T_n(\lambda)$ is

$$\mathrm{Var}(T_n(\lambda)) = \frac{a_n^2}{n} 2\pi \Bigg(\int_\lambda^{\lambda + a_n^{-1}} f_X(\omega)^2 d\omega$$

$$+ \int_\lambda^{\lambda + a_n^{-1}} \int_{-\pi}^{\lambda + a_n^{-1}} Q_X(\omega_1, \omega_2, -\omega_2) d\omega_1 d\omega_2$$

$$- \int_{-\pi}^{\lambda} \int_\lambda^{\lambda + a_n^{-1}} Q_X(\omega_1, \omega_2, -\omega_2) d\omega_1 d\omega_2 \Bigg) + \text{lower order.} \quad (3.12)$$

We can see the result from (3.12) by cases:

(i) if $a_n = n^\beta$ where $0 < \beta < 1$, then the limiting variance of $T_n(\lambda)$ is

$$\mathrm{Var}(T_n(\lambda)) \to 0,$$

(ii) if $a_n = n^\beta$ where $\beta > 1$, then the limiting variance of $T_n(\lambda)$ is

$$\mathrm{Var}(T_n(\lambda)) \to \infty,$$

(iii) if $a_n = n^\beta$ where $\beta = 1$, then the limiting variance of $T_n(\lambda)$ is

$$\mathrm{Var}(T_n(\lambda)) \to f_X(\lambda)^2.$$

Thus, the conclusion holds. □

Remark 3.2 The result in Lemma 3.1 seems surprising at first glance, since it may be expected that the asymptotic variance of (3.11) does not depend on the order of factor n^β. However, the phenomenon can be explained in a heuristic way. Returning back to the definition of $T_n(\lambda)$, the quantity

$$\frac{2\pi}{n} \sum_{\lambda < \omega_s \le \lambda + n^{-\beta}} I_{n,X}(\omega_s). \quad (3.13)$$

Looking at the number of periodograms $I_{n,X}(\omega_s)$ with different Fourier frequencies ω_s, we can find that (3.13) depends on the order of the length $n^{-\beta}$. If $0 < \beta < 1$, then more and more periodograms will be involved in the summation as n increases. Conversely, if $\beta > 1$, then the interval for the frequency will be much smaller as n increases. Only the case $\beta = 1$ keeps the same order between the number of periodograms and the length of the interval, and therefore only one periodogram $I_{n,X}(\omega_s)$ is involved in the summation.

Following Lemma 3.1, the proof of Theorem 3.5 is given as follows.

Proof (Theorem 3.5) Consider the following process:

$$M_n(\delta) = n \left\{ S_n\left(\lambda_p + \frac{\delta}{\sqrt{n}}\right) - S_n(\lambda_p) \right\}.$$

By Knight's identity (see (Knight 1998)), we have

$$
\begin{aligned}
M_n(\delta) &= -\delta\sqrt{n}\left\{\frac{2\pi}{n}\sum_{\omega_j\in\Lambda}\big(p-\mathbb{1}(\omega_j<\lambda_p)\big)\big(I_{n,X}(\omega_j)-f_X(\omega_j)\big)\right\} \\
&\quad + \frac{2\pi}{n}\sum_{\omega_j\in\Lambda}\left[\int_0^{\delta/\sqrt{n}}n\big(\mathbb{1}(\omega_j\le\lambda_p+s)-\mathbb{1}(\omega_j\le\lambda_p)\big)ds\right]I_{n,X}(\omega_j)+o(1) \\
&= -\delta\sqrt{n}\left\{\frac{2\pi}{n}\sum_{\omega_j\in\Lambda}\big(p-\mathbb{1}(\omega_j<\lambda_p)\big)\big(I_{n,X}(\omega_j)-f_X(\omega_j)\big)\right\} \\
&\quad + n\int_0^{\delta/\sqrt{n}}\left(\frac{2\pi}{n}\sum_{\lambda_p<\omega_j\le\lambda_p+s}I_{n,X}(\omega_j)\right)ds+o(1) \\
&= M_{n1}(\delta)+M_{n2}(\delta)+o(1),\quad\text{(say)}.
\end{aligned}
\tag{3.14}
$$

The asymptotic distribution of $\sqrt{n}(\hat{\lambda}_p-\lambda_p)$ can be derived from the asymptotic distribution of $M_n(\delta)$ by the argument of minimization. The asymptotic distribution of $M_n(\delta)$ is determined by Slutsky's theorem applying to the asymptotic joint distribution of $(M_{n1}(\delta),M_{n2}(\delta))^{\mathsf{T}}$, where $M_{n1}(\delta)$ is asymptotically normal and $M_{n2}(\delta)$ converges in probability.

Let us first consider $M_{n1}(\delta)$. From Theorem 7.6.1 in Brillinger (2001), it holds that

$$
M_{n1}(\delta)\xrightarrow{\mathcal{L}}\mathcal{N}(0,\underline{\sigma}_p^2),
$$

where $\underline{\sigma}_p^2$ is given by

$$
\begin{aligned}
\underline{\sigma}_p^2 &= 4\pi p^2\int_{-\pi}^{\pi}f_X(\omega)^2d\omega+2\pi(1-4p)\int_{-\pi}^{\lambda_p}f_X(\omega)^2d\omega+2\pi\int_{-\lambda_p}^{\lambda_p}f_X(\omega)^2d\omega \\
&\quad +2\pi\left\{\int_{-\pi}^{\lambda_p}\int_{-\pi}^{\lambda_p}Q_X(\omega_1,\omega_2,-\omega_2)d\omega_1d\omega_2\right. \\
&\quad +\int_{-\pi}^{\pi}\int_{-\pi}^{\pi}p^2Q_X(\omega_1,\omega_2,-\omega_2)d\omega_1d\omega_2 \\
&\quad \left. -2p\int_{-\pi}^{\lambda_p}\int_{-\pi}^{\pi}Q_X(\omega_1,\omega_2,-\omega_2)d\omega_1d\omega_2\right\}.
\end{aligned}
$$

Next, we consider $M_{n2}(\delta)$. From (3.12) in Lemma 3.1, it holds that for any sequence $a_n\to\infty$,

$$
\mathrm{Var}\left(\frac{2\pi}{n}\sum_{\lambda<\omega_j\le\lambda+a_n^{-1}}I_{n,X}(\omega_j)\right)=O(a_n^{-1}n^{-1}).
$$

For a large enough constant C, it holds that

$$\mathrm{Var}\left(\frac{2\pi}{n}\sum_{\lambda<\omega_j\leq\lambda+a_n^{-1}}I_{n,X}(\omega_j)\right)\leq Ca_n^{-1}n^{-1}.$$

Note that the integration interval for s in (3.14) is $[0,\delta/\sqrt{n}]$. Thus, $M_{n2}(\delta)$ can be evaluated by

$$\mathrm{Var}(M_{n2}(\delta))$$

$$= n^2\mathrm{Var}\left\{\int_0^{\delta/\sqrt{n}}\left(\frac{2\pi}{n}\sum_{\lambda_p<\omega_j\leq\lambda_p+s}I_{n,X}(\omega_j)\right)ds\right\}$$

$$= n^2\int_0^{\delta/\sqrt{n}}\int_0^{\delta/\sqrt{n}}\mathrm{Cov}\left(\sum_{\lambda_p<\omega_j\leq\lambda_p+s}I_{n,X}(\omega_j),\sum_{\lambda_p<\omega_k\leq\lambda_p+t}I_{n,X}(\omega_k)d\omega\right)dsdt$$

$$\leq n^2\int_0^{\delta/\sqrt{n}}\int_0^{\delta/\sqrt{n}}\mathrm{Var}\left(\sum_{\lambda_p<\omega_j\leq\lambda_p+s}I_{n,X}(\omega_j)\right)^{1/2}$$

$$\times\ \mathrm{Var}\left(\sum_{\lambda_p<\omega_k\leq\lambda_p+t}I_{n,X}(\omega_k)\right)^{1/2}dsdt$$

$$\leq n^2\int_0^{\delta/\sqrt{n}}\int_0^{\delta/\sqrt{n}}C^2s^{1/2}t^{1/2}n^{-1}dsdt$$

$$= C^2n\cdot\left(\frac{2}{3}\delta^{3/2}n^{-3/4}\right)^2\to 0.$$

By Chebyshev's inequality, it holds that

$$|M_{n2}(\delta)-E(M_{n2}(\delta))|\xrightarrow{\mathscr{P}}0. \tag{3.15}$$

The expectation of $M_{n2}(\delta)$ can be evaluated as follows. From Theorem 4.3.2 in Brillinger (2001),

$$E(M_{n2}(\delta)) = E\left(n \int_0^{\delta/\sqrt{n}} \left(\frac{2\pi}{n} \sum_{\lambda_p < \omega_j \le \lambda_p + s} I_{n,X}(\omega_j)\right) ds\right)$$

$$= n \int_0^{\delta/\sqrt{n}} E\left(\frac{2\pi}{n} \sum_{\lambda_p < \omega_j \le \lambda_p + s} I_{n,X}(\omega_j)\right) ds$$

$$= n \int_0^{\delta/\sqrt{n}} \left(\int_{\lambda_p}^{\lambda_p + s} f_X(\omega) d\omega + O(n^{-1})\right) ds$$

$$= \frac{1}{2\pi} \sum_{h \in \mathbb{Z}} R_X(h) \int_0^{\delta/\sqrt{n}} \left(\int_{\lambda_p}^{\lambda_p + s} n\, e^{-ih\omega} d\omega\right) ds + o(1).$$

Note that if we change ω by $\tilde{\omega} = \omega - \lambda_p$, then it holds that

$$\int_{\lambda_p}^{\lambda_p + s} n\, e^{-ih\omega} d\omega = e^{-ih\lambda_p} \int_0^s n\, e^{-ih\tilde{\omega}} d\tilde{\omega}.$$

Thus, we have

$$E(M_{n2}(\delta)) = \frac{1}{2\pi} \sum_{h \in \mathbb{Z}} R_X(h) e^{-ih\lambda_p} \int_0^{\delta/\sqrt{n}} \left(\int_0^s n\, e^{-ih\tilde{\omega}} d\tilde{\omega}\right) ds + o(1). \quad (3.16)$$

Changing the order of integration of double integrals in (3.16), it holds that

$$E(M_{n2}(\delta))$$
$$= \frac{1}{2\pi} \sum_{h \in \mathbb{Z}} R_X(h) e^{-ih\lambda_p} \int_0^{\delta/\sqrt{n}} \left(\int_{\tilde{\omega}}^{\delta/\sqrt{n}} n\, e^{-ih\tilde{\omega}} ds\right) d\tilde{\omega} + o(1)$$

$$= \frac{1}{2\pi} \sum_{h \in \mathbb{Z}} R_X(h) e^{-ih\lambda_p} \int_0^{\delta/\sqrt{n}} e^{-ih\tilde{\omega}} (\delta\sqrt{n} - n\tilde{\omega}) d\tilde{\omega} + o(1). \quad (3.17)$$

For any $h \in \mathbb{Z}$, by the definition of the derivative, as $n \to \infty$,

$$\int_0^{\delta/\sqrt{n}} \delta\sqrt{n}\, e^{-ih\tilde{\omega}} d\tilde{\omega} = \delta\left(\sqrt{n} \int_0^{\delta/\sqrt{n}} e^{-ih\tilde{\omega}} d\tilde{\omega}\right) \to \delta^2. \quad (3.18)$$

By L'Hospital's rule and the definition of the second derivative, we have, for any $h \in \mathbb{Z}$,

$$\int_0^{\delta/\sqrt{n}} n\tilde{\omega}\, e^{-ih\tilde{\omega}} d\tilde{\omega} \to \frac{1}{2}\delta^2 \frac{d}{d\tilde{\omega}}\left(\tilde{\omega} e^{-ih\tilde{\omega}}\right)\Big|_{\tilde{\omega}=0} = \frac{1}{2}\delta^2. \quad (3.19)$$

By (3.18), (3.19) and the definition of spectral density, (3.17) can be evaluated by

$$E(M_{n2}(\delta)) \to \frac{1}{2}\delta^2 f_X(\lambda_p). \tag{3.20}$$

Therefore, by (3.15) and (3.20) with the triangular inequality, it holds that

$$M_{n2}(\delta) \xrightarrow{\mathscr{P}} \frac{1}{2}\delta^2 f_X(\lambda_p).$$

Now, applying Slutsky's theorem to the joint distribution of $(M_{n1}(\delta), M_{n2}(\delta))^\mathsf{T}$, we obtain

$$M_n(\delta) \xrightarrow{\mathscr{L}} M(\delta) = -\delta\mathscr{N} + \frac{1}{2}\delta^2 f_X(\lambda_p), \tag{3.21}$$

where $\mathscr{N}$ has mean 0 and variance σ_p^2. Note that the right-hand side in Eq. (3.21) is minimized by $\delta = f_X(\lambda_p)^{-1}\mathscr{N}$. By Corollary 2 in Knight (1998), the desired result holds, i.e.,

$$\sqrt{n}(\hat{\lambda}_p - \lambda_p) \xrightarrow{\mathscr{L}} \mathscr{N}(0, f_X(\lambda_p)^{-2}\sigma_p^2).$$

This completes the proof.   □

3.4 Sinusoid Model

In this section, we consider a modified quantile estimator $\hat{\lambda}_p^*$ by smoothing for spectral quantiles in the frequency domain. Asymptotic normality of $\hat{\lambda}_p^*$ is shown under sinusoid models. Sinusoid models constitute a broader class than what we have considered in Sect. 3.3.

First, let us introduce the sinusoid model

$$Y(t) = \sum_{j=1}^{J} R_j \cos(\omega_j t + \phi_j) + X(t), \tag{3.22}$$

where $\{X(t)\}$ is a zero-mean second-order stationary process satisfying Assumption 3.4 as before. $\{\phi_j\}$ is uniformly distributed on $(-\pi, \pi)$, and is independent of $\{X(t)\}$. The amplitude $\{R_j\}$ and the frequency $\{\omega_j\}$ are real constants.

For the stochastic process $\{Y(t)\}$, the autocovariance function $R_Y(h)$ of $\{Y(t)\}$ is

$$R_Y(h) = \frac{1}{2}\sum_{j=1}^{J} R_j^2 \cos(\omega_j h) + R_X(h). \tag{3.23}$$

From (3.23), it is not difficult to see that $\{Y(t)\}$ is also second-order stationary. From (1.10), the spectral distribution function $F_Y(\omega)$ is represented by

$$F_Y(\omega) = \frac{1}{2} \sum_{j=1}^{J} R_j^2 \mathscr{H}(\omega - \omega_j) + F_X(\omega), \qquad (3.24)$$

where $\mathscr{H}(\omega)$ is so-called Heaviside step function such that

$$\mathscr{H}(\omega) = \begin{cases} 1, & \text{if } \omega \geq 0, \\ 0, & \text{otherwise.} \end{cases}$$

Accordingly, the spectral densify $f_Y(\omega)$ is written by the generalized derivative as

$$f_Y(\omega) = \frac{1}{2} \sum_{j=1}^{J} R_j^2 \delta(\omega - \omega_j) + f_X(\omega), \qquad (3.25)$$

where $\delta(\omega)$ is the Dirac delta function.

In the following, we provide two figures of spectral distributions of the process $\{Y(t)\}$. The frequency and the amplitude of one are $\omega_1 = \pi/2$ and $R_1 = 1/2$, and the other are $\omega_1 = \pi/6$ and $R_1 = 1/2$. The process $\{X(t)\}$ is supposed to be the MA(1) process with coefficient 0.9.

From Fig. 3.1, we see that the spectral distributions for the processes $\{X(t)\}$ and $\{Y(t)\}$ are almost the same. However, the spectral distributions for the process $\{Y(t)\}$ are not continuous at the specific frequency of the trigonometric function.

Let us introduce a modified quantile estimator $\hat{\lambda}_p^*$ for the spectral distribution function of the sinusoid model $\{Y(t)\}$. Let $I_{n,Y}^*(\omega)$ be

$$I_{n,Y}^*(\omega) = \sum_{|h|<n} C_n^Y(h) \exp(-ih\omega), \qquad (3.26)$$

where $C_n^Y(h)$ is the sample autocovariance of $\{Y(t)\}$. We use a spectral window $\phi(\omega)$ to smooth $I_{n,Y}^*(\omega)$ as follows. $w(x)$ is the corresponding window function. That is,

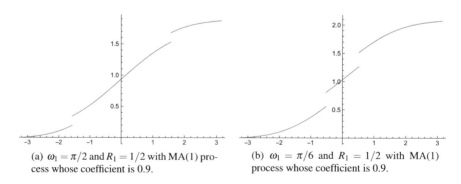

(a) $\omega_1 = \pi/2$ and $R_1 = 1/2$ with MA(1) process whose coefficient is 0.9.

(b) $\omega_1 = \pi/6$ and $R_1 = 1/2$ with MA(1) process whose coefficient is 0.9.

Fig. 3.1 Spectral distributions of Sinusoid models

$$\hat{f}_Y(\omega) = \frac{1}{2\pi} \int_{-\pi}^{\pi} \phi(\omega - \lambda) I_{n,Y}^*(\lambda) d\lambda$$

$$= \frac{1}{2\pi} \sum_{|h| \leq m} w\left(\frac{h}{m}\right) C_n^Y(h) e^{-ih\omega}.$$

Assumptions on the spectral window function $\phi(\omega)$ are given as follows.

Assumption 3.6 Let $\phi(\omega)$ satisfy

(i) $m \to \infty$, $m/n \to 0$ and $m^2/n \to \infty$, as $n \to \infty$.
(ii) $w(0) = 1$.
(iii) $w(-x) = w(x)$ and $|w(x)| \leq 1$ for all $x \in [-1, 1]$.
(iv) $w(x) = 0$ for $|x| > 1$.
(v) There exists a real constant c_1 such that

$$\int_{-\infty}^{\infty} \frac{1}{m} \phi\left(\frac{l}{m}\right) dl \to c_1,$$

as $m \to \infty$.
(vi) The pair (ϕ, f_Y) satisfies $\phi(\cdot) f_Y(\cdot) \in \mathscr{L}^u$ for some u, $1 < u \leq 2$, and suppose that there exists $c > 0$ such that

$$\sup_{|\lambda| < \varepsilon} \|\phi(\cdot)\{f_Y(\cdot) - f_Y(\cdot - \lambda)\}\|_u = O(\varepsilon^c)$$

as $\varepsilon \to 0$.

Here, $\mathscr{L}^p$ denotes the space of complex-valued functions on $[-\pi, \pi]$, equipped with L^p norm $\|g\|_p$, i.e., $\{\int_{-\pi}^{\pi} |g(\omega)|^p d\omega\}^{1/p}$.

Now, let us introduce a modified quantile estimator $\hat{\lambda}_p^*$. Let us define the objective function $S_n^*(\theta)$ as

$$S_n^*(\theta) = \int_{-\pi}^{\pi} \rho_p(\omega - \theta) \hat{f}_Y(\omega) d\omega.$$

The modified estimator $\hat{\lambda}_p^*$ is

$$\hat{\lambda}_p^* = \arg\min_{\theta \in \Lambda} S_n^*(\theta). \tag{3.27}$$

In addition, we add the following assumption for the quantile λ_p of the sinusoid model since the spectral distribution $F_Y(\omega)$ has jumps.

Assumption 3.7 The spectral distribution function $F_Y(\omega)$ has a density $f_Y(\omega)$ in a neighborhood of λ_p and $f_Y(\omega)$ is continuous at λ_p with $0 < f_Y(\lambda_p) < \infty$.

Theorem 3.8 *Suppose the process $\{Y(t) : t \in \mathbb{Z}\}$ is defined by (3.22) with $\{X(t)\}$ satisfying Assumption 3.2. The pth quantile λ_p satisfies Assumption 3.7. If $\hat{\lambda}_p^*$ is defined by (3.27) under Assumption 3.6, then we have*

$$\hat{\lambda}_p^* \xrightarrow{\mathscr{P}} \lambda_p.$$

To prove Theorem 3.8, we need the following lemma.

Lemma 3.2 *Assume $\sum_{j_1,j_2,j_3=-\infty}^{\infty} |\mathrm{cum}_X(j_1, j_2, j_3)| < \infty$. For any square-integrable function $\phi(\omega)$,*

$$\int_{-\pi}^{\pi} \left(I_{n,Y}(\omega) - E I_{n,Y}(\omega)\right)\phi(\omega)d\omega \xrightarrow{\mathscr{P}} 0. \tag{3.28}$$

Proof Let

$$\tilde{\phi}(n) = \frac{1}{2\pi} \int_{-\pi}^{\pi} \phi(\omega) \exp(in\omega)d\omega.$$

From Hosoya and Taniguchi (1982) and Li et al. (1994), it holds that

$$\mathrm{Var}\left\{ \int_{-\pi}^{\pi} (I_{n,Y}(\omega) - E I_{n,Y}(\omega))\phi(\omega)d\omega \right\} =$$

$$\frac{1}{n^2} \sum_{t_1,t_2,t_3,t_4=1}^{n} \tilde{\phi}(t_1 - t_2)\tilde{\phi}(t_3 - t_4)\Big\{ R_Y(t_3 - t_1)R_Y(t_4 - t_2)$$

$$+ R_Y(t_4 - t_1)R_Y(t_3 - t_2) + Q_Y(t_2 - t_1, t_3 - t_1, t_4 - t_1)\Big\}$$

$$= \frac{2\pi}{n} \int_{-\pi}^{\pi} (\phi(\omega)\overline{\phi(\omega)} + \phi(\omega)\overline{\phi(-\omega)}) f_Y(\omega) f_X(\omega)d\omega$$

$$+ \frac{2\pi}{n} \int_{-\pi}^{\pi} \int_{-\pi}^{\pi} \phi(\omega_1)\phi(-\omega_2)Q_X(\omega_1, \omega_2, -\omega_2)d\omega_1 d\omega_2.$$

Here, $f_Y(\omega)$ is defined in (3.25). From Chebyshev's inequality, (3.28) holds. $\qquad\square$

Now we provide the proof of Theorem 3.8.

Proof (Theorem 3.8) We only have to show the pointwise limit of $S_n^*(\theta)$ is $S(\theta)$. The rest of the argument for the proof is similar to the proof of Theorem 3.3. Note that $\hat{f}_Y(\omega)$ has a representation such that

$$\hat{f}_Y(\omega) = \int_{-\pi}^{\pi} \phi(\omega - \lambda)I_{n,Y}^*(\lambda)d\lambda.$$

Similarly, we have

$$|S_n^*(\theta) - S(\theta)| \le \left| \int_{-\pi}^{\pi} \rho_p(\omega - \theta)(\hat{f}_Y(\omega) - E\hat{f}_Y(\omega))d\omega \right|$$

$$+ \left| \int_{-\pi}^{\pi} \rho_p(\omega - \theta)E(\hat{f}_Y(\omega))d\omega - \int_{-\pi}^{\pi} \left(\int_{-\pi}^{\pi} \rho_p(\omega - \theta)\phi(\omega - \lambda)d\omega \right) F_Y(d\lambda) \right|.$$

The first term in the right-hand side converges to 0 in probability, which can be seen from Lemma 3.2. Under Assumption 3.6 (vi), we see that the second term in the right-hand side converges to 0 from Theorem 1.1 in Hosoya (1997). □

Next, we consider the asymptotic distribution of the modified estimator $\hat{\lambda}_p^*$.

Theorem 3.9 *Suppose $\{Y(t) : t \in \mathbb{Z}\}$ is defined by (3.22) with $\{X(t)\}$ satisfying Assumption 3.6. The pth quantile λ_p satisfies Assumption 3.7. If $\hat{\lambda}_p^*$ is defined by (3.27) under Assumption 3.6, then we have*

$$\sqrt{n}(\hat{\lambda}_p^* - \lambda_p) \xrightarrow{\mathcal{L}} \mathcal{N}(0, \sigma_p^{*2}),$$

where

$$\sigma_p^{*2} = f_Y(\lambda_p)^{-2}\left[4\pi p^2 \int_{-\pi}^{\pi} f_Y(\omega)f_X(\omega)d\omega \right.$$

$$+ 2\pi(1 - 4p)\int_{-\pi}^{\lambda_p} f_Y(\omega)f_X(\omega)d\omega + 2\pi \int_{-\lambda_p}^{\lambda_p} f_Y(\omega)f_X(\omega)d\omega$$

$$+ 2\pi \left\{ \int_{-\pi}^{\lambda_p} \int_{-\pi}^{\lambda_p} Q_X(\omega_1, \omega_2, -\omega_2)d\omega_1 d\omega_2 \right.$$

$$+ \int_{-\pi}^{\pi} \int_{-\pi}^{\pi} p^2 Q_X(\omega_1, \omega_2, -\omega_2)d\omega_1 d\omega_2$$

$$\left. \left. - 2p \int_{-\pi}^{\lambda_p} \int_{-\pi}^{\pi} Q_X(\omega_1, \omega_2, -\omega_2)d\omega_1 d\omega_2 \right\} \right].$$

Here, $f_Y(\omega)$ is the spectral density defined in (3.25).

Proof Consider the following process:

$$M_n^*(\delta) = n\left\{ S_n^*\left(\lambda_p - \frac{\delta}{\sqrt{n}} \right) - S_n^*(\lambda_p) \right\}.$$

By Knight's identity, we have

$$M_n^*(\delta) = -\delta\sqrt{n}\left\{ \int_{-\pi}^{\pi} (p - \mathbb{1}(\omega < \lambda_p))\hat{f}_Y(\omega)d\omega \right\}$$

$$+ \int_{-\pi}^{\pi} \int_0^{\delta/\sqrt{n}} n\left(\mathbb{1}(\omega \le \lambda_p + s) - \mathbb{1}(\omega \le \lambda_p) \right)\hat{f}_Y(\omega)ds d\omega$$

$$= M_{n1}^*(\delta) + M_{n2}^*(\delta), \quad \text{(say)}.$$

Let us first evaluate the term $M_{n1}^*(\delta)$. Note that $f_Y(\omega)$ is continuous at λ_p, it holds that

$$\int_{-\pi}^{\pi} (p - \mathbb{1}(\omega < \lambda_p)) f_Y(\omega) d\omega = 0.$$

Thus, the term $M_{n1}^*(\delta)$ can be written as

$$M_{n1}^*(\delta) = -\delta\sqrt{n}\left\{\int_{-\pi}^{\pi} (p - \mathbb{1}(\omega < \lambda_p))(\hat{f}_Y(\omega) - c_1 f_Y(\omega)) d\omega\right\}$$

$$= -\delta\sqrt{n}\left\{\int_{-\pi}^{\pi} (p - \mathbb{1}(\omega < \lambda_p))\right.$$

$$\left.\times\left((2\pi)^{-1}\int_{-\pi}^{\pi} \phi(\omega - \lambda)I_{n,Y}^*(\lambda) d\lambda - f_Y(\omega)\right) d\omega\right\}$$

$$= -\delta\sqrt{n}\left\{\int_{-\pi}^{\pi} (p - \mathbb{1}(\omega < \lambda_p))\right.$$

$$\times (2\pi)^{-1}\int_{-\pi}^{\pi} \phi(\omega - \lambda)\left(I_{n,Y}^*(\lambda) - f_Y(\lambda)\right) d\lambda\, d\omega$$

$$+\int_{-\pi}^{\pi}\left(p - \mathbb{1}(\omega < \lambda_p)\right)$$

$$\left.\times\left((2\pi)^{-1}\int_{-\pi}^{\pi} \phi(\omega - \lambda) f_Y(\lambda) d\lambda - c_1 f_Y(\omega)\right) d\omega\right\}$$

$$= -\delta\sqrt{n}(M_{n11}^* + M_{n12}^*), \quad \text{(say)}.$$

Now, focusing on the part of $I_{n,Y}^*(\lambda) - f_Y(\lambda)$ in the term M_{n11}^*, we change variables $(\omega, \lambda) \mapsto (\eta, \lambda)$ by setting $\eta = m(\omega - \lambda)$, so noting Assumption 3.6 (v), it holds that

$$M_{n11}^* = \int_{-\pi}^{\pi}\left[\int_{m(-\pi-\lambda)}^{m(\pi-\lambda)}\left(p - \mathbb{1}\left(\lambda + \frac{\eta}{m} < \lambda_p\right)\right)(2\pi m)^{-1}\phi\left(\frac{\eta}{m}\right) d\eta\right]$$

$$\times\left(I_{n,Y}^*(\lambda) - f_Y(\lambda)\right) d\lambda$$

$$= c_1\int_{-\pi}^{\pi}\left(p - \mathbb{1}\left(\lambda + \frac{\eta}{m} < \lambda_p\right)\right)\left(I_{n,Y}^*(\lambda) - f_Y(\lambda)\right) d\lambda$$

$$+\int_{-\pi}^{\pi}\left[\int_{m(-\pi-\lambda)}^{m(\pi-\lambda)}\left(p - \mathbb{1}\left(\lambda + \frac{\eta}{m} < \lambda_p\right)\right)(2\pi m)^{-1}\phi\left(\frac{\eta}{m}\right) d\eta\right.$$

$$\left.-c_1\left(p - \mathbb{1}\left(\lambda + \frac{\eta}{m} < \lambda_p\right)\right)\right]\left(I_{n,Y}^*(\lambda) - f_Y(\lambda)\right) d\lambda$$

$$= M_{n111}^* + M_{n112}^*, \quad \text{(say)}.$$

From Lemma A3.3 in Hosoya and Taniguchi (1982), we can see that

$$\sqrt{n}M_{n111}^* \xrightarrow{\mathcal{L}} c_1\mathcal{N}(0, \bar{\sigma}_p^2),$$

where $\bar{\sigma}_p^2$ is

$$
\bar{\sigma}_p^2 = 4\pi p^2 \int_{-\pi}^{\pi} f_Y(\omega) f_X(\omega) d\omega + 2\pi(1-4p) \int_{-\lambda_p}^{\lambda_p} f_Y(\omega) f_X(\omega) d\omega
$$

$$
+ 2\pi \int_{-\lambda_p}^{\lambda_p} f_Y(\omega) f_X(\omega) d\omega + 2\pi \Big\{ \int_{-\pi}^{\lambda_p} \int_{-\pi}^{\lambda_p} Q_X(\omega_1, \omega_2, -\omega_2) d\omega_1 d\omega_2
$$

$$
+ \int_{-\pi}^{\pi} \int_{-\pi}^{\pi} p^2 Q_X(\omega_1, \omega_2, -\omega_2) d\omega_1 d\omega_2
$$

$$
- 2p \int_{-\pi}^{\lambda_p} \int_{-\pi}^{\pi} Q_X(\omega_1, \omega_2, -\omega_2) d\omega_1 d\omega_2 \Big\}.
$$

On the other hand, the following term R in the square brackets in M_{n112}^* can be evaluated by the triangular inequality as follows:

$$
R = \int_{m(-\pi-\lambda)}^{m(\pi-\lambda)} \Big(p - \mathbb{1}\Big(\lambda + \frac{\eta}{m} < \lambda_p\Big)\Big)
$$

$$
\times (2\pi m)^{-1} \phi\Big(\frac{\eta}{m}\Big) d\eta - c_1\Big(p - \mathbb{1}\Big(\lambda + \frac{\eta}{m} < \lambda_p\Big)\Big)
$$

$$
\le \Big| \int_{m(-\pi-\lambda)}^{m(\pi-\lambda)} \Big(p - \mathbb{1}\Big(\lambda + \frac{\eta}{m} < \lambda_p\Big)\Big)(2\pi m)^{-1}\phi\Big(\frac{\eta}{m}\Big) d\eta
$$

$$
- \int_{m(-\pi-\lambda)}^{m(\pi-\lambda)} \Big(p - \mathbb{1}(\lambda < \lambda_p)\Big)(2\pi m)^{-1}\phi\Big(\frac{\eta}{m}\Big) d\eta \Big|
$$

$$
+ \Big| \int_{m(-\pi-\lambda)}^{m(\pi-\lambda)} \Big(p - \mathbb{1}(\lambda < \lambda_p)\Big)(2\pi m)^{-1}\phi\Big(\frac{\eta}{m}\Big) d\eta - c_1\Big(p - \mathbb{1}(\lambda < \lambda_p)\Big) \Big|.
$$

The first absolute difference can be evaluated by

$$
\Big| \int_{m(-\pi-\lambda)}^{m(\pi-\lambda)} \Big(p - \mathbb{1}\Big(\lambda + \frac{\eta}{m} < \lambda_p\Big)\Big)(2\pi m)^{-1}\phi\Big(\frac{\eta}{m}\Big) d\eta
$$

$$
- \int_{m(-\pi-\lambda)}^{m(\pi-\lambda)} \Big(p - \mathbb{1}(\lambda < \lambda_p)\Big)(2\pi m)^{-1}\phi\Big(\frac{\eta}{m}\Big) d\eta \Big|
$$

$$
\le \int_{m(-\pi-\lambda)}^{m(\pi-\lambda)} \Big| \mathbb{1}\Big(\lambda + \frac{\eta}{m} < \lambda_p\Big) - \mathbb{1}(\lambda < \lambda_p) \Big|(2\pi m)^{-1}\phi\Big(\frac{\eta}{m}\Big) d\eta
$$

$$
\le \int_{m(\lambda_p-\lambda)}^{m(\pi-\lambda)} \mathbb{1}\Big(\lambda < \lambda_p\Big)(2\pi m)^{-1}\phi\Big(\frac{\eta}{m}\Big) d\eta
$$

$$
+ \int_{m(-\pi-\lambda)}^{m(\lambda_p-\lambda)} \mathbb{1}(\lambda > \lambda_p)(2\pi m)^{-1}\phi\Big(\frac{\eta}{m}\Big) d\eta
$$

$$
\to 0,
$$

as $m \to \infty$, by the definition of Lebesgue integration. The second absolute difference can be shown to be asymptotically 0 by Assumption 3.6 (v).

Since the term $(\sqrt{n} M^*_{n112})^2$ is uniformly integrable by Assumption 3.4, we see that by Vitali's convergence theorem, $\mathrm{Var}(\sqrt{n} M^*_{n112}) \to 0$, as $n \to \infty$. In turn, it holds that $\sqrt{n} M^*_{n112} \xrightarrow{\mathscr{P}} 0$.

Next, we evaluate the term M^*_{n12} in $M^*_{n1}(\delta)$. From Theorem 10 in Hannan (1970), page 283, we see, by Assumption 3.6, that

$$\left| (2\pi)^{-1} \int_{-\pi}^{\pi} \phi(\omega - \lambda) f_Y(\lambda) d\lambda - c_1 f_Y(\omega) \right| = O(m^{-1}).$$

Therefore, $\sqrt{n} M^*_{n12} = O(n^{1/2}/m)$, which shows that $\sqrt{n} M^*_{n12} \to 0$ under Assumption 3.6 (i). In summary, it holds that $M^*_{n1}(\delta) \xrightarrow{\mathscr{L}} -\delta c_1 \mathcal{N}(0, \bar{\sigma}^2_p)$.

As for the second term $M^*_{n2}(\delta)$, we have

$$M^*_{n2}(\delta) = \int_0^{\delta/\sqrt{n}} \int_{\lambda_p}^{\lambda_p + s} n \, \hat{f}_Y(\omega) d\omega ds$$

$$= \int_0^{\delta\sqrt{n}} \left(\int_{\lambda_p}^{\lambda_p + t/n} \hat{f}_Y(\omega) d\omega \right) dt.$$

Now, note that $\mathrm{Var}(\hat{f}_Y(\omega)) = O(m/n)$ uniformly in $\omega \in \Lambda$ under Assumption 3.6. With similar computation in the proof of Theorem 3, we can see that

$$M^*_{n2}(\delta) \xrightarrow{\mathscr{P}} \frac{1}{2} \delta^2 c_1 f_Y(\lambda_p).$$

Thus, by Slutsky's theorem, we obtain

$$M^*_n(\delta) \xrightarrow{\mathscr{L}} M^*(\delta) = -\delta c_1 \mathcal{N} + \frac{1}{2} \delta^2 c_1 f_Y(\lambda_p),$$

which is minimized by $\delta = f_Y(\lambda_p)^{-1} \mathcal{N}$. Therefore,

$$\sqrt{n}(\hat{\lambda}_p - \lambda_p) \xrightarrow{\mathscr{L}} \mathcal{N}(0, f_Y(\lambda_p)^{-2} \bar{\sigma}^2_p),$$

and the asymptotic variance σ^{*2}_p in Theorem 3.9 is $\sigma^{*2}_p = f_Y(\lambda_p)^{-2} \bar{\sigma}^2_p$. This completes the proof. □

From Theorems 3.3 and 3.9, we see that both quantile estimators $\hat{\lambda}_p$ and $\hat{\lambda}^*_p$ have asymptotic normality. Taking it into account, we consider the quantile test in the frequency domain in the next section.

3.5 Quantile Test

In this section, we consider quantile tests in the frequency domain. The hypothesis testing problem is

$$H : \lambda_p = q \quad \text{versus} \quad A : \lambda_p \neq q. \tag{3.29}$$

Let $\tilde{\lambda}_p$ be $\hat{\lambda}_p$ or $\hat{\lambda}_p^*$ and $\tilde{\sigma}_p^2$ be σ_p^2 or σ_p^{*2}. For the testing problem (3.29), we define the test statistic T_n as

$$T_n = \frac{\tilde{\lambda}_p - q}{\tilde{\sigma}_p}. \tag{3.30}$$

As what we considered in Sect. 3.2, T_n is applicable to discriminate second-order stationary processes. One commonly considers testing problem for second-order stationary process under Assumption 3.2, which guarantees the continuity of the spectral density function. Therefore, our test statistic T_n has a broader range of applications than the common one. It is possible to choose a proper quantile or multiple quantiles as our interest to implement the hypothesis testing.

Now, let us provide the asymptotic distribution of the test statistic T_n, which is a corollary of Theorems 3.5.

Corollary 3.1 *Suppose the process $\{X(t)\}$ satisfies Assumption 3.4 or the process $\{Y(t)\}$ is defined by (3.22) with $\{X(t)\}$ satisfying Assumption 3.4. The pth quantile λ_p satisfies Assumption 3.7. $\tilde{\lambda}_p$ is defined by (3.4) or (3.27) under Assumption 3.6. Then for test statistic T_n, defined by (3.30), we have*

(i) Under the null hypothesis H, $\sqrt{n}\, T_n \xrightarrow{\mathscr{L}} \mathscr{N}(0, 1)$;
(ii) Under the alternative hypothesis A, $\sqrt{n}\, (T_n - \frac{\lambda_p - q}{\tilde{\sigma}_p}) \xrightarrow{\mathscr{L}} \mathscr{N}(0, 1)$.

According to the significance level α, the hypothesis H is rejected if $\sqrt{n}\, |T_n| > \Phi_{1-\alpha/2}$, where $\Phi_{1-\alpha/2}$ is the $1 - \alpha/2$ percentage point of a standard normal distribution.

Remark 3.3 In practice, $\tilde{\sigma}_p$ is unknown. We have to estimate it in advance. It is possible to construct a consistent estimator for $\tilde{\sigma}_p$. (For higher order cumulant estimation, see Taniguchi 1982 or Keenan 1987.)

3.6 Numerical Studies

In this section, we implement the numerical studies to confirm the theoretical results in Sects. 3.3 and 3.4.

3.6.1   Finite Sample Performance

In this subsection, we first investigate the finite sample performance of the quantile estimator $\hat{\lambda}_p$ with different numbers of observations $n = 100, 1000$. The model considered here is the Gaussian AR(1) model with coefficient -0.9. The standard deviations σ of innovation processes are $\sigma = 1$. p in this simulation is set by $p = 0.7$.

In Fig. 3.2, we provide the Q–Q plots of $\sqrt{n}(\hat{\lambda}_{0.7} - \lambda_{0.7})$ when $n = 100$ and $n = 1000$ in 200 simulations, respectively. It can be seen that the quantity is much closer to the normal distribution when the sample size is as large as $n = 1000$. In other words, the quantity may have a problem of approximating tails by normal distribution when the sample size is small.

Also, this problem happens when the standard deviation of innovation processes is small. In the following, we show the Q–Q plots of $\sqrt{n}(\hat{\lambda}_{0.7} - \lambda_{0.7})$ for the Gaussian white noise model and the Gaussian AR(1) model with coefficient -0.9, with the standard deviations $\sigma = 0.08$. $\hat{\lambda}_p$ is estimated from $n = 1000$ observations. The estimation procedures are repeated for 200 times to generate Q–Q plots.

In Fig. 3.3, we can see even the observations $n = 1000$, the tail of $\sqrt{n}(\hat{\lambda}_{0.7} - \lambda_{0.7})$ is not well captured by normal distribution when $\sigma = 0.08$ for both second-order

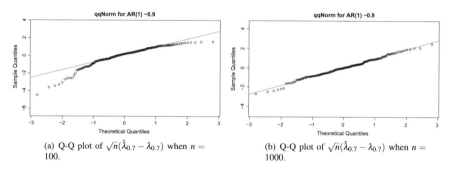

(a) Q-Q plot of $\sqrt{n}(\hat{\lambda}_{0.7} - \lambda_{0.7})$ when $n = 100$.

(b) Q-Q plot of $\sqrt{n}(\hat{\lambda}_{0.7} - \lambda_{0.7})$ when $n = 1000$.

Fig. 3.2   Q–Q plots of $\sqrt{n}(\hat{\lambda}_{0.7} - \lambda_{0.7})$ with observations $n = 100$ and $n = 1000$

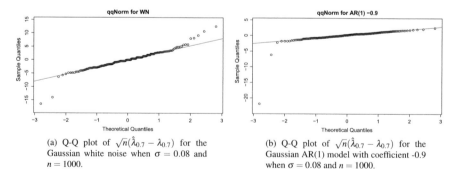

(a) Q-Q plot of $\sqrt{n}(\hat{\lambda}_{0.7} - \lambda_{0.7})$ for the Gaussian white noise when $\sigma = 0.08$ and $n = 1000$.

(b) Q-Q plot of $\sqrt{n}(\hat{\lambda}_{0.7} - \lambda_{0.7})$ for the Gaussian AR(1) model with coefficient -0.9 when $\sigma = 0.08$ and $n = 1000$.

Fig. 3.3   Q–Q plots of $\sqrt{n}(\hat{\lambda}_{0.7} - \lambda_{0.7})$ for the Gaussian white noise models

Table 3.1 The unbiased sample variance of $\sqrt{n}\hat{\lambda}_{0.7}$ for the Gaussian white noise models

	Standard deviation of innovation process				
n	$\sigma = 0.2$	$\sigma = 0.4$	$\sigma = 0.6$	$\sigma = 0.8$	$\sigma = 1.0$
200	4.675	4.538	4.569	4.453	5.088
400	5.372	4.652	4.689	4.785	4.516
600	5.594	5.093	4.772	4.665	4.711
800	6.300	5.833	5.007	5.038	4.702
1000	5.821	5.799	5.107	4.649	5.042

Table 3.2 The unbiased sample variance of $\sqrt{n}\hat{\lambda}_{0.7}$ for the Gaussian AR(1) models with coefficient 0.9

	Standard deviation of innovation process				
n	$\sigma = 0.2$	$\sigma = 0.4$	$\sigma = 0.6$	$\sigma = 0.8$	$\sigma = 1.0$
200	0.379	0.362	0.337	0.358	0.343
400	0.441	0.324	0.387	0.303	0.348
600	0.369	0.363	0.356	0.351	0.356
800	0.353	0.376	0.349	0.393	0.394
1000	0.390	0.339	0.344	0.374	0.366

stationary models. The extreme value of $\sqrt{n}(\hat{\lambda}_p - \lambda_p)$ sometimes could be very large compared to the tail of a normal distribution.

Furthermore, to see the variance of quantile estimator $\hat{\lambda}_p$ in finite sample case, we generate the following unbiased sample variance of $\sqrt{n}\hat{\lambda}_{0.7}$ for the Gaussian white noise models and the Gaussian AR(1) models with coefficient 0.9 when observations $n = 200, 400, 600, 800$, and 1000 by 1000 simulations in Tables 3.1 and 3.2. The standard deviations of innovation processes are set to be $\sigma = 0.2, 0.4, 0.6, 0.8$, and 1.0.

3.6.2 Numerical Results for Estimation

In this subsection, we first investigate the performance of the estimator $\hat{\lambda}_p$ defined by (3.4). As a benchmark, we fix the standard deviations of innovation processes as follows. The stochastic process considered here are second-order stationary processes, including the Gaussian white noise model, the Gaussian MA(1) model with coefficient 0.9, the Gaussian AR(1) model with coefficient 0.9, and the Gaussian AR(1) model with coefficient -0.9, whose standard deviations σ of innovation processes are set to be $\sigma = 1$. The spectral distribution functions for these four models are already given in Fig. 1.1. The dependence structures of them are obviously different.

In Tables 3.3, 3.4, 3.5 and 3.6, we summarized the true value of λ_p, the mean and the root mean squared deviation (RMSD) of the quantile estimator $\hat{\lambda}_p$ for the spectral distribution function of each model by 30, 50, and 100 observations for 1000 simulations. We set p as $p = 0.6, 0.7, 0.8, 0.9$.

From Tables 3.3, 3.4, 3.5 and 3.6, we can see that the quantile estimator $\hat{\lambda}_{p,n}$ has consistency for the true λ_p. As the size of observations becomes larger, the bias between the estimator and the true value becomes smaller. RMSD also becomes smaller as the observations increase.

Next, we compare the performance of $\hat{\lambda}_p^*$ with $\hat{\lambda}_p$. We consider the same models, i.e., the Gaussian white noise model, the Gaussian MA(1) model with coefficient 0.9, the Gaussian AR(1) model with coefficient 0.9 and the Gaussian AR(1) model with coefficient -0.9, whose standard deviations σ of innovation processes are set to be $\sigma = 1$. For brevity, denote each model by WN, MA, AR1, and AR2, respectively.

Table 3.3 The true value of λ_p, the mean (RMSD) of the estimated quantiles $\hat{\lambda}_{p,n}$ of the spectral distribution of White noise model for 1000 simulations

p	White noise			
	λ_p	$\hat{\lambda}_{p,30}$	$\hat{\lambda}_{p,50}$	$\hat{\lambda}_{p,100}$
0.6	0.628	0.698 (0.284)	0.665 (0.233)	0.649 (0.170)
0.7	1.257	1.272 (0.370)	1.273 (0.289)	1.267 (0.208)
0.8	1.885	1.861 (0.368)	1.863 (0.294)	1.879 (0.211)
0.9	2.513	2.444 (0.297)	2.473 (0.225)	2.500 (0.168)

Table 3.4 The true value of λ_p, the mean (RMSD) of the estimated quantiles $\hat{\lambda}_{p,n}$ of the spectral distribution of MA(1) model with coefficient 0.9 for 1000 simulations

p	MA(1)			
	λ_p	$\hat{\lambda}_{p,30}$	$\hat{\lambda}_{p,50}$	$\hat{\lambda}_{p,100}$
0.6	0.318	0.388 (0.187)	0.364 (0.154)	0.337 (0.120)
0.7	0.653	0.705 (0.254)	0.682 (0.201)	0.667 (0.152)
0.8	1.032	1.067 (0.278)	1.038 (0.220)	1.037 (0.160)
0.9	1.520	1.545 (0.265)	1.528 (0.218)	1.513 (0.156)

Table 3.5 The true value of λ_p, the mean (RMSD) of the estimated quantiles $\hat{\lambda}_{p,n}$ of the spectral distribution of AR(1) model with coefficient 0.9 for 1000 simulations

p	AR(1) with coefficient 0.9			
	λ_p	$\hat{\lambda}_{p,30}$	$\hat{\lambda}_{p,50}$	$\hat{\lambda}_{p,100}$
0.6	0.034	0.099 (0.077)	0.075 (0.054)	0.058 (0.038)
0.7	0.076	0.178 (0.126)	0.136 (0.095)	0.108 (0.061)
0.8	0.145	0.304 (0.213)	0.237 (0.143)	0.192 (0.096)
0.9	0.321	0.604 (0.359)	0.501 (0.278)	0.414 (0.194)

Table 3.6 The true value of λ_p, the mean (RMSD) of the estimated quantiles $\hat{\lambda}_{p,n}$ of the spectral distribution of AR(1) model with coefficient -0.9 for 1000 simulations

p	λ_p	$\hat{\lambda}_{p,30}$	$\hat{\lambda}_{p,50}$	$\hat{\lambda}_{p,100}$
			AR(1) with coefficient -0.9	
0.6	2.820	2.563 (0.359)	2.656 (0.266)	2.732 (0.178)
0.7	2.997	2.834 (0.198)	2.909 (0.138)	2.950 (0.095)
0.8	3.065	2.963 (0.124)	3.003 (0.104)	3.034 (0.057)
0.9	3.107	3.046 (0.079)	3.067 (0.056)	3.081 (0.039)

Table 3.7 The true value of λ_p, the mean (RMSD) of the estimated quantiles $\hat{\lambda}_{0.6,100}$, and the mean (RMSD) of the estimated quantiles $\hat{\lambda}^*_{0.6,100}$ with bandwidth $m = 20, 50, 80$ for spectral distribution in each model for 1000 simulations

Model	$\lambda_{0.6}$	$\hat{\lambda}_{0.6,100}$	$\hat{\lambda}^*_{0.6,100}$ with bandwidth m		
			$m = 20$	$m = 50$	$m = 80$
WN	0.628	0.679 (0.175)	0.663 (0.142)	0.665 (0.153)	0.667 (0.157)
WN*	0.785	0.737 (0.189)	0.733 (0.153)	0.744 (0.171)	0.741 (0.179)
WN**	1.257	0.970 (0.360)	0.911 (0.378)	0.947 (0.357)	0.959 (0.353)
MA	0.318	0.364 (0.128)	0.355 (0.092)	0.346 (0.102)	0.351 (0.110)
MA*	0.362	0.370 (0.127)	0.397 (0.102)	0.392 (0.117)	0.389 (0.122)
MA**	0.499	0.439 (0.150)	0.466 (0.107)	0.461 (0.124)	0.461 (0.130)
AR1	0.034	0.088 (0.067)	0.095 (0.068)	0.095 (0.069)	0.091 (0.066)
AR1*	0.036	0.092 (0.070)	0.111 (0.079)	0.089 (0.064)	0.089 (0.066)
AR1**	0.041	0.097 (0.072)	0.121 (0.086)	0.096 (0.068)	0.095 (0.069)
AR2	2.820	2.737 (0.202)	2.569 (0.313)	2.634 (0.269)	2.666 (0.252)
AR2*	2.742	2.669 (0.254)	2.381 (0.419)	2.553 (0.293)	2.596 (0.273)
AR2**	1.925	2.381 (0.620)	2.092 (0.336)	2.253 (0.496)	2.295 (0.544)

In addition, we include all these models in the sinusoid model (3.22).

(i) Let $J = 1$. R_1 and ω_1 are set to be 1/2 and $\pi/2$. All these sinusoid models are denoted by WN*, MA*, AR1*, and AR2*, respectively.

(ii) Let $J = 1$. R_1 and ω_1 are set to be 1 and $\pi/2$. All these sinusoid models are denoted by WN**, MA**, AR1**, and AR2**, respectively.

We fix $p = 0.6$ in this simulation. $\hat{\lambda}_p$ and $\hat{\lambda}^*_p$ are calculated from 100 observations of each model. We use the Bartlett window as the lag window for the estimator $\hat{\lambda}^*_p$. We investigate the performance when the bandwidth m is $m = 20, 50, 80$ for 1000 simulations. The numerical results are shown in Table 3.7.

From Table 3.7, one can see that the modified estimator $\hat{\lambda}^*_{0.6}$ is closer to the true value and has smaller RMSD than the quantile estimator $\hat{\lambda}_{0.6}$ in the models WN and MA. On the other hand, the quantile estimator $\hat{\lambda}_{0.6}$ outperforms the modified estimator when the underlying model is AR1 or AR2. An adequate bandwidth could make the modified estimator $\hat{\lambda}^*$ better than $\hat{\lambda}$ in views of bias and RMSD but the choice of an adequate bandwidth in practice is a difficult problem, which can be found from this simulation study.

For the sinusoid models, the modified estimator $\hat{\lambda}^*_{0.6}$ has a better performance than the estimator $\hat{\lambda}_{0.6}$ in the sense of RMSD. The estimator $\hat{\lambda}_{0.6}$ sometimes has a lower bias than $\hat{\lambda}^*_{0.6}$ partly because there is a decrease in the autocorrelations of the trigonometric function for finite samples. The remarkable case in all the models is AR2**. In this case, the true quantile is close to the point of the jump in the spectral distribution. The modified estimator $\hat{\lambda}^*_{0.6}$ with any bandwidth is much better than $\hat{\lambda}_{0.6}$ in regard to bias and RMSE. We found that this property holds in general. In particular, from (3.24) we can see that if the amplitudes R_j become larger, then the jumps become the dominant part in the spectral distribution. In the case, the modified estimator $\hat{\lambda}^*$ becomes a good alternative to the estimator $\hat{\lambda}$.

Chapter 4
Empirical Likelihood Method for Time Series

Abstract Empirical likelihood method is one of the nonparametric statistical methods, which is applied to the hypothesis testing or construction of confidence regions for unknown parameters. This method has been developed for the statistical inference for independent and identically distributed random variables. To handle serial correlation, an empirical likelihood method is proposed in the frequency domain for second-order stationary processes. The Whittle likelihood is used as an estimating function in the empirical likelihood. It has been shown that the likelihood ratio test statistic based on the empirical likelihood is asymptotically χ^2-distributed. We discuss the application of the empirical likelihood method to symmetric α-stable linear processes. It is shown that the asymptotic distribution of our test statistic is quite different from the usual one. We illustrate the theoretical result with some numerical simulations.

4.1 Introduction

Not only parametric methods, nonparametric methods also have been developed for the statistical inference for time series models. A remarkable nonparametric method for the hypothesis testing and construction of confidence regions is called *empirical likelihood*. Owen (1988) introduced the empirical likelihood method to test the pivotal quantities for independent and identically distributed (i.i.d.) data. He showed that the empirical likelihood ratio statistic is asymptotically χ^2-distributed. Afterward, the empirical likelihood is linked to the estimating function approach in Qin and Lawless (1994). An extension to the dependent data in the time domain is discussed in Kitamura (1997).

For the approach in the frequency domain, Monti (1997) discussed the empirical likelihood for the Whittle estimation and derived the asymptotic distribution of the empirical likelihood ratio statistic. Ogata and Taniguchi (2010) developed the empirical likelihood approach for a class of vector-valued non-Gaussian stationary processes. The empirical likelihood is applied to the symmetric α-stable (sαs) processes in Akashi et al. (2015).

Y. Liu et al., *Empirical Likelihood and Quantile Methods for Time Series*,
JSS Research Series in Statistics, https://doi.org/10.1007/978-981-10-0152-9_4

4.2 Empirical Likelihood in the Frequency Domain

In this section, we describe the empirical likelihood and the estimating function from the Whittle likelihood in the frequency domain. Let us consider the following linear model:

$$X(t) = \sum_{j=0}^{\infty} \psi_j \varepsilon(t - j), \quad t \in \mathbb{Z}, \tag{4.1}$$

where $\psi_0 = 1$ and $\{\varepsilon(t) : t \in \mathbb{Z}\}$ is a sequence of i.i.d. random variables with variance σ^2. In addition, suppose the process (4.1) has the spectral density $g(\omega)$.

Let us review the h-step ahead linear prediction of a scalar stationary process $\{X(t) : t \in \mathbb{Z}\}$. An h-step ahead linear prediction problem is to predict $X(t)$ by a linear combination of $\{X(s) : s \le t - h\}$. Denote by $\hat{X}(t)$ the predictor and then

$$\hat{X}(t) = \sum_{j=h}^{\infty} \phi_j(\theta) X(t - j),$$

where $\{\phi_j(\theta)\}$ is the coefficients of the prediction, driven by a parameter $\theta \in \Theta \subset \mathbb{R}^d$, since it is unknown in advance. From Theorem 1.2, the spectral representation of the process $X(t)$ is

$$X(t) = \int_{-\pi}^{\pi} \exp(-it\omega) d\zeta_X(\omega),$$

and the spectral representation of the predictor $\hat{X}(t)$ is

$$\hat{X}(t) = \int_{-\pi}^{\pi} \exp(-it\omega) \sum_{j=h}^{\infty} \phi_j(\theta) \exp(ij\omega) d\zeta_X(\omega),$$

where $\{\zeta_X(\omega); -\pi \le \omega \le \pi\}$ is an orthogonal increment process satisfying

$$E[d\zeta_X(\omega) d\zeta_X(\mu)] = \begin{cases} g(\omega) d\omega & (\omega = \mu) \\ 0 & (\omega \ne \mu) \end{cases}.$$

Then, the prediction error is

$$E|X(t) - \hat{X}(t)|^2 = \int_{-\pi}^{\pi} \left| 1 - \sum_{j=h}^{\infty} \phi_j(\theta) \exp(ij\omega) \right|^2 g(\omega) d\omega. \tag{4.2}$$

The true parameter θ_0 is the minimizer of the prediction error (4.2). Especially, let us define $f(\omega; \theta)$ as

$$f(\omega; \theta) = \left| 1 - \sum_{j=h}^{\infty} \phi_j(\theta) \exp(ij\omega) \right|^{-2}.$$

Then the true parameter θ_0 is the minimizer of

$$\int_{-\pi}^{\pi} f(\omega; \theta)^{-1} g(\omega) d\omega. \tag{4.3}$$

A confidence region for the parameter θ_0 can be constructed from the Whittle likelihood if we have a nonparametric estimator for $g(\omega)$ in (4.3). The procedure mimics the parameter estimation method in Chap. 2. Thus, we introduce the empirical likelihood ratio as the testing statistic as follows.

Let $I_{n,X}(\omega)$ denote the periodogram of the observations, that is,

$$I_{n,X}(\omega) = \frac{1}{2\pi n} \left| \sum_{t=1}^{n} X(t) e^{it\omega} \right|^2.$$

The empirical likelihood ratio statistic is defined as

$$R(\theta) = \max_{w_1,\ldots,w_n} \left\{ \prod_{t=1}^{n} n w_t ; \sum_{t=1}^{n} w_t \, m(\lambda_t; \theta) = 0, \sum_{t=1}^{n} w_t = 1, \, 0 \le w_1, w_2, \ldots, w_n \le 1 \right\}, \tag{4.4}$$

where the estimating function $m(\lambda_t; \theta)$ in $R(\theta)$ is defined as

$$m(\lambda_t; \theta) = \frac{\partial}{\partial \theta} \frac{I_{n,X}(\lambda_t)}{f(\lambda_t; \theta)} \in \mathbb{R}^d, \quad \lambda_t = \frac{2\pi t}{n} \in (-\pi, \pi). \tag{4.5}$$

Suppose we are interested in the testing problem

$$H : \theta = \theta_0.$$

Generally, the estimating function m of the parameter θ for the empirical likelihood (4.4) is supposed to satisfy

$$E[\, m(\theta_0)\,] = 0, \qquad \theta_0 \in \Theta. \tag{4.6}$$

In our case, the true parameter θ_0 is the minimizer of (4.3) and thus the estimating function (4.5) satisfies (4.6). The product $\prod w_t$ can be regarded as an empirical version of the likelihood when the estimating equation $\sum_{t=1}^{n} w_t \, m(\lambda_t; \theta) = 0$ holds. The empirical likelihood ratio $R(\theta)$ is the ratio of the empirical likelihood to the likelihood $(1/n)^n$.

Now we review the asymptotic results for the process $\{X(t)\}$ with finite variance innovations. Let us assume the following assumptions for the process $\{X(t)\}$ and the parametric model $f(\theta)$.

Assumption 4.1 The process $\{X(t)\}$ is a zero-mean, strictly stationary real-valued process, all of whose moments exist. In addition,

(i) it holds that

$$\sum_{u_1,\ldots,u_{k-1}=-\infty}^{\infty} \left(1 + \sum_{j=1}^{k-1} |u_j|\right) |\text{cum}_X(u_1,\ldots,u_{k-1})| < \infty, \quad \text{for } k = 2, 3, \ldots,$$

where $\text{cum}_X(u_1,\ldots,u_{k-1})$ denotes the joint cumulant of $\big(X(t), X(t+u_1), \ldots, X(t+u_{k-1})\big)$;

(ii) let

$$C_k = \sum_{u_1,\ldots,u_{k-1}=-\infty}^{\infty} |\text{cum}_X(u_1,\ldots,u_{k-1})| < \infty$$

and it holds that

$$\sum_{k=1}^{\infty} \frac{C_k z^k}{k!} < \infty$$

for any z in a neighborhood of 0.

Let $\mathscr{F}(\Theta)$ be the family of parametric models.

Assumption 4.2 (i) The parametric models are defined as

$$\mathscr{F}(\Theta) = \left\{ g : g(\omega) = \sigma^2 \left| \sum_{j=0}^{\infty} g_j(\theta) \exp(-ij\omega) \right|^2 / (2\pi) \right\}.$$

(ii) The parameter space Θ is a compact subset of $\mathbb{R}^d$.
(iii) The parametric spectral density $f_\theta(\lambda)$ is twice continuously differentiable with respect to θ.

Theorem 4.3 *(Ogata and Taniguchi 2010) Under Assumptions 4.1 and 4.2, if the hypothesis $H : \theta = \theta_0$ holds, then we have*

$$-2 \log R(\theta_0) \xrightarrow{\mathscr{L}} \chi_d^2.$$

as $n \to \infty$, where χ_d^2 denotes a chi-square distribution with d degrees of freedom.

Proof See Ogata and Taniguchi (2010).

From Theorem 4.3, it is also possible to a construct confidence region for the parameters $\theta \in \Theta$. Actually, if one needs to construct a confidence region at the significant level q, then one can define the region as

$$C_{q,n} = \{\theta \in \Theta : -2 \log R(\theta) < c_q\},$$

where c_q is the $1 - q$ percentage point of a chi-square distribution with d degrees of freedom.

4.3 Empirical Likelihood for Symmetric α-stable Processes

In this section, we discuss the empirical likelihood method for sαs processes. We consider the same linear model

$$X(t) = \sum_{j=0}^{\infty} \psi_j Z(t - j), \quad t \in \mathbb{Z}. \tag{4.7}$$

but $\{Z(t) : t \in \mathbb{Z}\}$ is a sequence of i.i.d. sαs random variables with scale parameter $\sigma > 0$. The characteristic function of the random variables is expressed as

$$E \exp\{i\xi Z(1)\} = \exp\{-\sigma |\xi|^{\alpha}\}, \quad \xi \in \mathbb{R}.$$

In the case of $\alpha = 2$, the process is Gaussian. On the other hand, when α is less than 2, the usual spectral density $g(\omega)$ is not well defined. Here, we assume that $\alpha \in [1, 2)$ to guarantee probability convergence of important terms which will appear in proofs later.

Now we state the notations. For any sequence $\{A(t) : t \in \mathbb{Z}\}$ of random variables, let

$$\gamma_{n,A}^2 = n^{-2/\alpha} \sum_{t=1}^{n} A(t)^2,$$

$$I_{n,A}(\omega) = n^{-2/\alpha} \left| \sum_{t=1}^{n} A(t) \exp(it\omega) \right|^2,$$

$$\tilde{A}_t = \frac{A(t)}{\left(A(1)^2 + \cdots + A(n)^2\right)^{1/2}}, \quad t = 1, \cdots, n, \tag{4.8}$$

and

$$\tilde{I}_{n,A}(\omega) = \frac{I_{n,A}(\omega)}{\gamma_{n,A}^2} = \left| \sum_{t=1}^{n} \tilde{A}_t \exp(it\omega) \right|^2.$$

$\tilde{I}_{n,A}(\omega)$ is called a *self-normalized periodogram* of random variables $A(1), \cdots, A(n)$. The process $\{X(t)\}$ in (4.7) is well defined if we impose the following assumption.

Assumption 4.4 For some δ satisfying $0 < \delta < 1$,

$$\sum_{j=0}^{\infty} |j| |\psi_j|^{\delta} < \infty.$$

Under this assumption, the series of the right-hand side in (4.7) converges almost surely. This is an easy consequence of the three-series theorem (c.f. Petrov 1975). Furthermore, the process (4.7) has the normalized power transfer function

$$\tilde{g}(\omega) = \frac{1}{\psi^2} \left| \sum_{j=0}^{\infty} \psi_j \exp(ij\omega) \right|^2, \quad \psi^2 = \sum_{j=0}^{\infty} \psi_j^2.$$

From the property of stable random variables,

$$X(t) =_d \left\{ \sum_{j=0}^{\infty} |\psi_j|^{\alpha} \right\}^{1/\alpha} Z(1),$$

which implies that the process $\{X(t)\}$ does not have the finite second moment when $\alpha < 2$, so we cannot use the method of moments to make the statistical inference for the process. The empirical likelihood approach is still useful when we deal with the process $\{X(t)\}$. As discussed in Sect. 4.2, we define the parameter θ_0 of interest as the solution in

$$\frac{\partial}{\partial \theta} \int_{-\pi}^{\pi} \frac{\tilde{g}(\omega)}{f(\omega; \theta)} d\omega \bigg|_{\theta=\theta_0} = 0, \tag{4.9}$$

where $\theta \in \Theta \subset \mathbb{R}^d$. It is easy to understand the correspondence between (4.9) and (4.3). In the following, we give an example of the parameter θ_0 of interest.

Example 4.1 Let the model $f(\omega; \theta)$ be defined as

$$f(\omega; \theta) = |1 - \theta \exp(il\omega)|^{-2},$$

for fixed $l \in \mathbb{N}$. Then the left-hand side of (4.9) is

$$\frac{\partial}{\partial \theta} \int_{-\pi}^{\pi} \frac{\tilde{g}(\omega)}{f(\omega; \theta)} d\omega = \frac{\partial}{\partial \theta} \int_{-\pi}^{\pi} |1 - \theta \exp(il\omega)|^2 \left\{ \frac{1}{\psi^2} \left| \sum_{j=0}^{\infty} \psi_j \exp(ij\omega) \right|^2 \right\} d\omega$$

$$= \frac{1}{\psi^2} \sum_{j=0}^{\infty} \sum_{k=0}^{\infty} \int_{-\pi}^{\pi} (2\theta - 2\cos(l\omega)) \psi_j \psi_k \exp(i(j-k)\omega) d\omega$$

$$= \frac{1}{\psi^2} \left((4\pi)\theta \sum_{j=0}^{\infty} \psi_j^2 - (4\pi) \sum_{j=0}^{\infty} \psi_j \psi_{j+l} \right). \tag{4.10}$$

The solution of Eq. (4.10) is

$$\theta_0 = \frac{\sum_{j=0}^{\infty} \psi_j \psi_{j+l}}{\sum_{j=0}^{\infty} \psi_j^2},\tag{4.11}$$

where the right-hand side of (4.11) is the autocorrelation function of $s\alpha s$ process $\{X(t)\}$. From Davis and Resnick (1986), the sample autocorrelation function $\hat{\rho}(l)$, i.e.,

$$\hat{\rho}(l) = \frac{\sum_{t=1}^{n-l} X(t)X(t+l)}{\sum_{t=1}^{n} X(t)^2},\tag{4.12}$$

is consistent to the autocorrelation function θ_0 even for the $s\alpha s$ process. That is,

$$\hat{\rho}(l) \xrightarrow{\mathscr{P}} \theta_0.$$

The consistency of (4.12) in Example 4.1 is crucial. It motivates us to use the self-normalized periodogram $\tilde{I}_{n,X}(\omega)$ instead of the ordinary periodogram $I_{n,X}(\omega)$, since the self-normalized periodogram can be regarded as a natural transformation from the sample autocorrelation function $\hat{\rho}(l)$.

Let us define the empirical likelihood ratio statistic

$$R(\theta) = \max_{w_1,\ldots,w_n} \left\{ \prod_{t=1}^{n} n w_t ; \sum_{t=1}^{n} w_t m(\lambda_t; \theta) = 0, \sum_{t=1}^{n} w_t = 1, 0 \le w_1, w_2, \ldots, w_n \le 1 \right\},\tag{4.13}$$

with the following estimating function $m(\lambda_t; \theta)$:

$$m(\lambda_t; \theta) = \frac{\partial}{\partial \theta} \frac{\tilde{I}_{n,X}(\lambda_t)}{f(\lambda_t; \theta)} \in \mathbb{R}^d, \quad \lambda_t = \frac{2\pi t}{n} \in (-\pi, \pi).\tag{4.14}$$

Note that the estimating function $m(\lambda_t; \theta)$ of (4.5) and (4.14) shows the main difference between the empirical likelihood method for the processes with finite variance innovations and infinite variance innovations. In addition to Assumption 4.2, we impose the following assumption.

Assumption 4.5 For some $\mu \in (0, \alpha)$ and all $k = 1, \cdots, d$,

$$\sum_{t=1}^{\infty} \left| \int_{-\pi}^{\pi} \frac{\partial}{\partial \theta_k} \frac{\tilde{g}(\omega)}{f(\omega; \theta)} \Big|_{\theta=\theta_0} \cos(t\omega) d\omega \right|^{\mu} < \infty.$$

Assumption 4.5 is proposed in Klüppelberg and Mikosch (1996) for the $s\alpha s$ processes. It is not so strong since the AR(p) stable processes satisfying Assumption 4.1 also satisfy this assumption.

Let us investigate the asymptotic properties of the empirical likelihood ratio statistic $R(\theta)$ in (4.13) for the $s\alpha s$ processes (4.7). For brevity, let x_n be the normalizing sequence as

$$x_n = \left(\frac{n}{\log n}\right)^{1/\alpha}, \quad n = 2, 3, \cdots .$$

The asymptotic distribution of $R(\theta_0)$ under the null hypothesis $H : \theta = \theta_0$ is expressed in the following theorem.

Theorem 4.6 *Let $\alpha \in [1, 2)$. Suppose that Assumptions 4.2, 4.4, and 4.5 hold. Under the null hypothesis $H : \theta = \theta_0$, we have*

$$-\frac{2x_n^2}{n} \log R(\theta_0) \xrightarrow{\mathscr{L}} V^{\mathsf{T}} W^{-1} V, \qquad (4.15)$$

where V and W are $q \times 1$ random vector and $q \times q$ constant matrix, respectively. The jth and (k, l)th element of the vector V and the matrix W are expressed as

$$V_j = \frac{1}{\pi} \sum_{t=1}^{\infty} \frac{S_t}{S_0} \left\{ \int_{-\pi}^{\pi} \frac{\partial f(\omega; \theta)^{-1}}{\partial \theta_j} \bigg|_{\theta=\theta_0} \tilde{g}(\omega) \cos(t\omega) d\omega \right\},$$

$$W_{kl} = \frac{1}{2\pi} \int_{-\pi}^{\pi} \frac{\partial f(\omega; \theta)^{-1}}{\partial \theta_k} \frac{\partial f(\omega; \theta)^{-1}}{\partial \theta_l} \bigg|_{\theta=\theta_0} 2\tilde{g}(\omega)^2 d\omega$$

with independent random variables $S_0, S_1, S_2, \cdots$. Here, S_0 is a positive $\alpha/2$-stable random variable and $\{S_j : j = 1, 2, \cdots\}$ is a sequence of symmetric α-stable random variables.

To prove Theorem 4.6, we need the following preparation. For the sequence of random variables $\{A(t) : t \in \mathbb{Z}\}$ and $h = 1, \cdots, n - 1$, let $\rho_{n,A}(h)$ be

$$\rho_{n,A}(h) = \sum_{t=1}^{n-h} \tilde{A}_t \tilde{A}_{t+h},$$

where $\tilde{A}_t$ is defined in (4.8). Let $T_{n,A}(\omega)$ be

$$T_{n,A}(\omega) = 2 \sum_{h=1}^{n-1} \rho_{n,A}(h) \cos(h\omega).$$

Lemma 4.1 *Let $\alpha \in (0, 2]$. For the sequence of i.i.d. symmetric α-stable random variables $\{Z(t)\}$, the following results hold.*

(i) $ET_{n,Z}(\omega) = 0$,
(ii) *As $n \to \infty$, it holds that*

$$ET_{n,Z}(\omega)^2 \rightarrow \begin{cases} 1, & if \ \omega \neq 0 \quad \mod \pi, \\ 2, & if \ \omega = 0 \quad \mod \pi. \end{cases} \tag{4.16}$$

Proof We first prepare for the asymptotic moments of the normalized random variable $\tilde{Z}_t$. Note that the normalized random variable $\tilde{Z}_t$ is bounded almost surely. $E\tilde{Z}_t$ exists and is equal to 0 since $Z(t)$ is symmetric. From the definition of $\tilde{Z}_1, \tilde{Z}_2, \cdots, \tilde{Z}_n$, it always holds that

$$\sum_{t=1}^{n} \tilde{Z}_t^2 = 1 \quad \text{a.s..} \tag{4.17}$$

In addition, $\tilde{Z}_1, \tilde{Z}_2, \cdots, \tilde{Z}_n$ are identically distributed and thus $E\tilde{Z}_1^2 = 1/n$. Using Chebyshev's inequality, we can see

$$\Pr\left\{|\tilde{Z}_1| < \varepsilon^{-1/2}n^{-1/2}\right\} > 1 - \varepsilon$$

for any $\varepsilon > 0$. Thus we have $\sqrt{n}\,\tilde{Z}_1^2 = O_p(n^{-1/2})$ and $\sqrt{n}\,\tilde{Z}_1^2$ converges to 0 in probability. Therefore, by Taylor's theorem, there exists a constant c such that

$$E \exp\{i\xi \sqrt{n}\,\tilde{Z}_1^2\} = 1 - \frac{\xi^2}{2}nE\tilde{Z}_1^4 + \frac{\xi^3 \sin(\xi c)}{6}n^{3/2}E\tilde{Z}_1^6 + i\,\text{Im}\left[E \exp\{i\xi \sqrt{n}\,\tilde{Z}_1^2\}\right]. \tag{4.18}$$

On the other hand, by Lévy's continuity theorem, it holds that

$$E \exp\{i\xi \sqrt{n}\,\tilde{Z}_1^2\} \rightarrow 1. \tag{4.19}$$

From (4.18) and (4.19), we can conclude that

$$nE\tilde{Z}_1^4 \rightarrow 0 \tag{4.20}$$

as $n \rightarrow \infty$. Note that from (4.17), it holds that

$$\sum_{t=1}^{n} \tilde{Z}_t^4 + \sum_{t \neq s} \tilde{Z}_t^2 \tilde{Z}_s^2 = 1 \quad \text{a.s.} \tag{4.21}$$

Taking expectations on both sides of (4.21), we have

$$n(n-1)E\tilde{Z}_1^2\tilde{Z}_2^2 \rightarrow 1, \tag{4.22}$$

as $n \rightarrow \infty$.

(i) Now let us evaluate $ET_{n,Z}(\omega)$.

$$ET_{n,Z}(\omega) = 2E \sum_{h=1}^{n-1} \rho_{n,Z}(h) \cos(h\omega)$$

$$= 2E \sum_{h=1}^{n-1} \sum_{t=1}^{n-h} \tilde{Z}_t \tilde{Z}_{t+h} \cos(h\omega)$$

$$= 0,$$

since $Z(1), Z(2), \cdots, Z(n)$ are symmetric and independent.

(ii) Next, let us evaluate $T_{n,Z}(\omega)^2$. It holds that

$$ET_{n,Z}(\omega)^2 = n(n-1)E\tilde{Z}_1^2\tilde{Z}_2^2 + 2nE\tilde{Z}_1^2\tilde{Z}_2^2 \sum_{h=1}^{n-1} \cos(2h\omega) - 2E\tilde{Z}_1^2\tilde{Z}_2^2 \sum_{h=1}^{n-1} h \cos(2h\omega).$$

$$(4.23)$$

The first term of the right-hand side of (4.23) converges to 1 as $n \to \infty$ from (4.22). If $\omega = 0 \mod \pi$, then

$$2nE\tilde{Z}_1^2\tilde{Z}_2^2 \sum_{h=1}^{n-1} \cos(2h\omega) - 2E\tilde{Z}_1^2\tilde{Z}_2^2 \sum_{h=1}^{n-1} h \cos(2h\omega) = n(n-1)E\tilde{Z}_1^2\tilde{Z}_2^2 \to 1.$$

If $\omega \neq 0 \mod \pi$, we use the following two identical equations:

$$\sum_{h=1}^{n-1} \cos(2h\omega) = \frac{\cos(2(n-1)\omega) + \cos(2\omega) - \cos(2n\omega)}{2(1 - \cos(2\omega))}, \tag{4.24}$$

$$\sum_{h=1}^{n-1} h \cos(2h\omega) = \frac{n \cos(2(n-1)\omega) - (n-1)\cos(2n\omega) - 1}{2(1 - \cos(2\omega))}. \tag{4.25}$$

From Eqs. (4.24) and (4.25), we obtain that

$$2nE\tilde{Z}_1^2\tilde{Z}_2^2 \sum_{h=1}^{n-1} \cos(2h\omega) - 2E\tilde{Z}_1^2\tilde{Z}_2^2 \sum_{h=1}^{n-1} h \cos(2h\omega) \to 0.$$

Hence we conclude (4.16). □

Next, we evaluate the covariance of the self-normalized periodogram at the different Fourier frequencies.

Lemma 4.2 $\sum\sum_{k \neq l} \mathrm{Cov}\{\tilde{I}_{n,Z}(\lambda_k)^2, \tilde{I}_{n,Z}(\lambda_l)^2\} = O(n)$.

Proof Let $d_{n,Z}(\lambda_k) = \sum_{t=1}^{n} \tilde{Z}_t \exp(it\lambda_k)$. From Brillinger (2001),

$$\mathrm{Cov}\{\tilde{I}_{n,Z}(\lambda_k)^2, \tilde{I}_{n,Z}(\lambda_l)^2\} = \sum_{\nu:p=1}^{8} \prod_{j=1}^{p} \mathrm{cum}\{d_{n,Z}(\lambda_{k_j}); k_j \in \nu_j\},$$

where the summation is taken over all indecomposable partitions $\boldsymbol{v} = v_1 \cup \cdots \cup v_p$, $p = 1, \cdots, 8$, of the following table.

$$
\begin{array}{c|c|c}
k & k & -k & -k \\
\hline
l & l & -l & -l
\end{array}
\tag{4.26}
$$

Note that $\operatorname{cum}\{d_{n,Z}(\lambda_{k_1}), \cdots, d_{n,Z}(\lambda_{k_m})\} = 0$ for odd m.

For even m, let us first consider partitions for $p = 4$. We can evaluate the second-order cumulant as

$$
\begin{aligned}
\operatorname{cum}\{d_{n,Z}(\lambda_k), d_{n,Z}(\lambda_l)\} &= E\tilde{Z}_1^2 \sum_{t=1}^{n} \exp\bigl(it(\lambda_k - \lambda_l)\bigr) \\
&= \frac{1}{n} \sum_{t=1}^{n} \exp\left(it\frac{2\pi(k-l)}{n}\right) \\
&= \begin{cases} 1, & (k - l = 0 \mod n), \\ 0, & (k - l \neq 0 \mod n). \end{cases}
\end{aligned}
\tag{4.27}
$$

From (4.27), the partitions for $p = 4$ can be evaluated as

$$
\begin{aligned}
&\operatorname{cum}\{d_{n,Z}(\lambda_{k_1}), d_{n,Z}(\lambda_{k_2})\} \cdots \operatorname{cum}\{d_{n,Z}(\lambda_{k_7}), d_{n,Z}(\lambda_{k_8})\} \\
&= \begin{cases} 1, & (k_1 - k_2, \cdots, k_7 - k_8 \equiv 0 \mod n) \\ 0, & \text{otherwise.} \end{cases}
\end{aligned}
$$

Thus, when $p = 4$, it holds that

$$
\sum_{k \neq l} \prod_{j=1}^{4} \operatorname{cum}\{d_{n,Z}(\lambda_{k_j}); k_j \in v_j\} = O(n)
\tag{4.28}
$$

for any indecomposable partition of (4.26).

Next, we consider the following five types of partitions:

$$
\begin{aligned}
p = 1: \quad & (k, k, -k, -k, l, l, -l, -l), \\
p = 2: \quad & (k, -k, l, -l) \cup (k, -k, l, -l), \\
& (k, -k) \cup (k, -k, l, l, -l, -l), \\
& (l, -l) \cup (k, k, -k, -k, l, -l) \\
p = 3: \quad & (k, -k) \cup (l, -l) \cup (k, -k, l, -l).
\end{aligned}
\tag{4.29}
$$

We introduce generic residual terms $R_n^{(1)}(k, l)$, $R_n^{(2)}(k, l)$, and $R_n^{(3)}(k, l)$ as follows. The fourth-order joint cumulant on $(\lambda_k, -\lambda_k, \lambda_l, -\lambda_l)$ is represented as

$$\mathrm{cum}\{d_{n,Z}(\lambda_k), d_{n,Z}(-\lambda_k), d_{n,Z}(\lambda_l), d_{n,Z}(-\lambda_l)\}$$
$$= nE\tilde{Z}_1^4 + n(n-1)E\tilde{Z}_1^2\tilde{Z}_2^2 - 1 + R_n^{(1)}(k,l). \qquad (4.30)$$

From (4.20) and (4.22), (4.30) becomes

$$\mathrm{cum}\{d_{n,Z}(\lambda_k), d_{n,Z}(-\lambda_k), d_{n,Z}(\lambda_l), d_{n,Z}(-\lambda_l)\} = R_n^{(1)}(k,l) + o(1). \qquad (4.31)$$

Note that the following identities hold:

$$\left(\sum_{t=1}^{n}\tilde{Z}_t^2\right)\left(\sum_{t,s}^{(*)}\tilde{Z}_t^2\tilde{Z}_s^2\right) = \sum_{t,s}^{(*)}\tilde{Z}_t^2\tilde{Z}_s^2 = 2\sum_{t,s}^{(*)}\tilde{Z}_t^4\tilde{Z}_s^2 + \sum_{t,s,u}^{(*)}\tilde{Z}_t^2\tilde{Z}_s^2\tilde{Z}_u^2,$$
$$\left(\sum_{t=1}^{n}\tilde{Z}_t^2\right)\left(\sum_{t=1}^{n}\tilde{Z}_t^4\right) = \sum_{t=1}^{n}\tilde{Z}_t^4 = \sum_{t,s}^{(*)}\tilde{Z}_t^4\tilde{Z}_s^2 + \sum_{t=1}^{n}\tilde{Z}_t^6.$$

From (4.17), we have

$$\sum_{t=1}^{n}\tilde{Z}_t^8 + 4\sum_{t,s}^{(*)}\tilde{Z}_t^6\tilde{Z}_s^2 + 3\sum_{t,s}^{(*)}\tilde{Z}_t^4\tilde{Z}_s^4$$
$$+ 6\sum_{t,s,u}^{(*)}\tilde{Z}_t^4\tilde{Z}_s^2\tilde{Z}_u^2 + \sum_{t,s,u,v}^{(*)}\tilde{Z}_t^2\tilde{Z}_s^2\tilde{Z}_u^2\tilde{Z}_v^2 = 1, \qquad \text{a.s.,}$$

where $\sum_{t_1,\cdots,t_m}^{(*)}$ is a summation taken over all $t_1, \cdots, t_m$, which are different from each other. By the similar argument as above, we obtain that

$$\mathrm{cum}\{d_{n,Z}(\lambda_k), d_{n,Z}(-\lambda_k), d_{n,Z}(\lambda_l), d_{n,Z}(\lambda_l), d_{n,Z}(-\lambda_l), d_{n,Z}(-\lambda_l)\}$$
$$= R_n^{(2)}(k,l) + o(1), \qquad (4.32)$$

and the eighth-order joint cumulant as

$$\mathrm{cum}\{d_{n,Z}(\lambda_k), d_{n,Z}(-\lambda_k), d_{n,Z}(\lambda_k), d_{n,Z}(-\lambda_k),$$
$$d_{n,Z}(\lambda_l), d_{n,Z}(\lambda_l), d_{n,Z}(-\lambda_l), d_{n,Z}(-\lambda_l)\}$$
$$= 2n^2 E\tilde{Z}_1^4\tilde{Z}_2^4 - 6n^3 E\tilde{Z}_1^4\tilde{Z}_2^2\tilde{Z}_3^2$$
$$+ n^4 E\tilde{Z}_1^2\tilde{Z}_2^2\tilde{Z}_3^2\tilde{Z}_4^2 - \left(n^2 E\tilde{Z}_1^2\tilde{Z}_2^2\right)^2 + R_n^{(3)}(k,l) + o(1), \qquad (4.33)$$

With a similar argument as (4.27), it is possible to show that $\sum\sum_{k\neq l} R_n^{(\eta)}(k,l) = O(n)$ for $\eta = 1, 2, 3$. From (4.28), (4.29), (4.31), (4.32), and (4.33), we see that

$$\sum_{k\neq l}\sum_{\mathbf{v}:p=1}^{8}\prod_{j=1}^{p}\mathrm{cum}\{d_{n,Z}(\lambda_{k_j}); k_j \in v_j\} = O(n).$$

We omit the detailed proof for the reader.   □

Let $P_n(\theta_0)$ and $S_n(\theta_0)$ be defined as follows.

$$P_n(\theta_0) = \frac{1}{n} \sum_{t=1}^{n} m(\lambda_t; \theta_0) \quad \text{and} \quad S_n(\theta_0) = \frac{1}{n} \sum_{t=1}^{n} m(\lambda_t; \theta_0) m(\lambda_t; \theta_0)^{\mathsf{T}}.$$

Lemma 4.3 *Under Assumption 4.2,*

$$S_n(\theta_0) \xrightarrow{\mathscr{P}} W$$

as $n \to \infty$. Here W is defined in Theorem 3.1.

Proof We first make use of the decomposition of the periodogram in Klüppelberg and Mikosch (1994) as follows, i.e.,

$$\tilde{I}_{n,X}(\omega)^2 = \tilde{g}(\omega)^2 \tilde{I}_{n,Z}(\omega)^2 + o_p(1) \tag{4.34}$$

$$= \tilde{g}(\omega)^2 \left\{ 1 + 2 \sum_{h=1}^{n-1} \rho_{n,Z}(h) \cos(h\omega) \right\}^2 + o_p(1)$$

$$= \tilde{g}(\omega)^2 \{ 1 + 2T_{n,Z}(\omega) + T_{n,Z}(\omega)^2 \} + o_p(1).$$

From Lemma 4.1, we obtain that

$$E\left[S_n(\theta_0)\right] = \frac{1}{n} \sum_{t=1}^{n} \left. \frac{\partial f(\lambda_t; \theta)^{-1}}{\partial \theta} \frac{\partial f(\lambda_t; \theta)^{-1}}{\partial \theta^{\mathsf{T}}} \right|_{\theta=\theta_0} E\tilde{I}_{n,X}(\lambda_t)^2$$

$$\to \frac{1}{2\pi} \int_{-\pi}^{\pi} \left. \frac{\partial f(\omega; \theta)^{-1}}{\partial \theta} \frac{\partial f(\omega; \theta)^{-1}}{\partial \theta^{\mathsf{T}}} \right|_{\theta=\theta_0} 2\tilde{g}(\omega)^2 d\omega = W.$$

Let us define $h_{\theta_0}(\omega)_{jk}$ as

$$h_{\theta_0}(\omega)_{jk} = \left. \frac{\partial f(\omega; \theta)^{-1}}{\partial \theta_a} \frac{\partial f(\omega; \theta)^{-1}}{\partial \theta_b} \right|_{\theta=\theta_0} \tilde{g}(\omega)^2.$$

From Lemma 4.2, Assumption 4.2, and (4.34), we have

$$\text{Cov}\big(S_n(\theta_0)_{jk}, S_n(\theta_0)_{lm}\big)$$

$$= \frac{1}{n^2} \sum_{t=1}^{n} \sum_{s=1}^{n} h_{\theta_0}(\lambda_t)_{jk} h_{\theta_0}(\lambda_s)_{lm} \text{Cov}\big(\tilde{I}_{n,Z}(\lambda_t)^2, \tilde{I}_{n,Z}(\lambda_s)^2\big)$$

$$= \frac{1}{n^2} \sum_{t=1}^{n} h_{\theta_0}(\lambda_t)_{jk} h_{\theta_0}(\lambda_t)_{lm} \text{Var}\big(\tilde{I}_{n,Z}(\lambda_t)^2\big)$$

$$+ \frac{1}{n^2} \sum_{t \neq s} h_{\theta_0}(\lambda_t)_{jk} h_{\theta_0}(\lambda_s)_{lm} \text{Cov}\big(\tilde{I}_{n,Z}(\lambda_t)^2, \tilde{I}_{n,Z}(\lambda_s)^2\big) + o(1)$$

$$\to 0$$

for $j, k, l, m = 1, \cdots, d$. The result implies the convergence of $S_n(\theta_0)$ in probability. $\qquad\square$

In the following, we prove Theorem 4.6.

Proof (Theorem 4.6) By Lagrange's multiplier method, the weights $w_1, \cdots, w_n$ maximizing the objective function in $R(\theta)$ are given by

$$w_t = \frac{1}{n} \frac{1}{1 + \phi^{\mathsf{T}} m(\lambda_t; \theta_0)}, \quad t = 1, \cdots, n,$$

where $\phi \in \mathbb{R}^d$ is the Lagrange multiplier which is defined as the solution of d-restrictions

$$J_{n,\theta_0}(\phi) = \frac{1}{n} \sum_{t=1}^{n} \frac{m(\lambda_t; \theta_0)}{1 + \phi^{\mathsf{T}} m(\lambda_t; \theta_0)} = 0. \qquad (4.35)$$

First of all, let us derive the asymptotic order of ϕ. Let Y_t be

$$Y_t = \phi^{\mathsf{T}} m(\lambda_t; \theta_0).$$

From (4.35), we have

$$0 = \frac{1}{n} \sum_{t=1}^{n} \frac{m(\lambda_t; \theta_0)}{1 + Y_t}$$

$$= \frac{1}{n} \sum_{t=1}^{n} \left\{ 1 - Y_t + \frac{Y_t^2}{1 + Y_t} \right\} m(\lambda_t; \theta_0) \qquad (4.36)$$

$$= P_n(\theta_0) - S_n(\theta_0)\phi + \frac{1}{n} \sum_{t=1}^{n} \frac{m(\lambda_t; \theta_0) Y_t^2}{1 + Y_t}.$$

Hence, ϕ can be expressed as

$$\phi = S_n(\theta_0)^{-1}\left\{P_n(\theta_0) + \frac{1}{n}\sum_{t=1}^{n}\frac{m(\lambda_t;\theta_0)Y_t^2}{1+Y_t}\right\}$$

$$\equiv S_n(\theta_0)^{-1}P_n(\theta_0) + \varepsilon \quad \text{(say)}. \tag{4.37}$$

Let M_n be $M_n = \max_{1\leq k\leq n}\|m(\lambda_k;\theta_0)\|$. Under Assumption 4.2, M_n can be evaluated by

$$M_n = \max_{1\leq t\leq n}\left\|\frac{\partial f(\lambda_t;\theta)^{-1}}{\partial\theta}\right|_{\theta=\theta_0}\tilde{I}_{n,X}(\lambda_t)\right\|$$

$$\leq \max_{1\leq t\leq n}\left\|\frac{\partial f(\lambda_t;\theta)^{-1}}{\partial\theta}\right|_{\theta=\theta_0}\right\|\max_{1\leq t\leq n}|I_{n,X}(\lambda_t)|\frac{1}{\gamma_{n,X}^2}$$

$$\leq \max_{\omega\in[-\pi,\pi]}\left\|\frac{\partial f(\omega;\theta)^{-1}}{\partial\theta}\right|_{\theta=\theta_0}\right\|\max_{\omega\in[-\pi,\pi]}|I_{n,X}(\omega)|\frac{1}{\gamma_{n,X}^2}$$

$$= \max_{\omega\in[-\pi,\pi]}\left\|\frac{\partial f(\omega;\theta)^{-1}}{\partial\theta}\right|_{\theta=\theta_0}\right\|\max_{\omega\in[-\pi,\pi]}|g(\omega)|\frac{\max_{\omega\in[-\pi,\pi]}\|I_{n,X}(\omega)\|}{\max_{\omega\in[-\pi,\pi]}|g(\omega)|}\frac{1}{\gamma_{n,X}^2}$$

$$\leq \max_{\omega\in[-\pi,\pi]}\left\|\frac{\partial f(\omega;\theta)^{-1}}{\partial\theta}\right|_{\theta=\theta_0}\right\|\max_{\omega\in[-\pi,\pi]}|g(\omega)|\max_{\omega\in[-\pi,\pi]}\left|\frac{I_{n,X}(\omega)}{g(\omega)}\right|\frac{1}{\gamma_{n,X}^2}$$

$$= C\max_{\omega\in[-\pi,\pi]}\left|\frac{I_{n,X}(\omega)}{g(\omega)}\right|,$$

where C is a generic constant. By Assumption 4.4 and Corollary 3.3 of Mikosch et al. (2000), we obtain

$$M_n = O_p(\beta_n^2),$$

where

$$\beta_n = \begin{cases} (\log n)^{1-1/\alpha}, & \text{when } 1 < \alpha < 2, \\ \log\log n, & \text{when } \alpha = 1. \end{cases}$$

Here, we discuss the case of $1 < \alpha < 2$. The case of $\alpha = 1$ can be considered by a parallel argument. From Ogata and Taniguchi (2010), there exists a unit vector $\boldsymbol{u}$ in $\mathbb{R}^d$ such that $\phi = \|\phi\|\boldsymbol{u}$ and

$$\|\phi\|\left\{\boldsymbol{u}^\top S_n(\theta_0)\boldsymbol{u} - \boldsymbol{u}^\top M_n P_n(\theta_0)\right\} \leq \boldsymbol{u}^\top P_n(\theta_0). \tag{4.38}$$

Actually, note that $\boldsymbol{u}^\top J_{n,\theta_0}(\phi) = 0$. From (4.36), it also holds that

$$\boldsymbol{u}^\top P_n(\theta_0) = \boldsymbol{u}^\top\left(\frac{1}{n}\sum_{t=1}^{n}\frac{\phi^\top m(\lambda_t;\theta_0)}{1+Y_t}m(\lambda_t;\theta_0)\right)$$

and by $\phi = \|\phi\|\boldsymbol{u}$, we have

$$\|\phi\|\boldsymbol{u}^{\mathsf{T}}\boldsymbol{S}_n(\theta_0)(1 + Y_t)^{-1}\boldsymbol{u} = \boldsymbol{u}^{\mathsf{T}}\boldsymbol{P}_n(\theta_0). \qquad (4.39)$$

By $(1 + Y_t)^{-1} = 1 - Y_t/(1 + Y_t)$, we obtain (4.38) from (4.39).

To determine the order of the term ϕ, we first discuss the asymptotics of the term $\boldsymbol{P}_n(\theta_0)$. Lemma P5.1 of Brillinger (2001) allows us to write $x_n \boldsymbol{P}_n(\theta_0)$ as

$$
\begin{aligned}
x_n \boldsymbol{P}_n(\theta_0) &= \frac{1}{2\pi} x_n \int_{-\pi}^{\pi} \frac{\partial f(\omega; \theta)}{\partial \theta}\bigg|_{\theta=\theta_0} \tilde{I}_{n,X}(\omega) d\omega + O_p\left(\frac{x_n}{n}\right) \\
&= \frac{1}{2\pi} \frac{1}{\gamma_{n,X}^2} x_n \int_{-\pi}^{\pi} \frac{\partial f(\omega; \theta)}{\partial \theta}\bigg|_{\theta=\theta_0} \left\{ I_{n,X}(\omega) - T_n \psi^2 \tilde{g}(\omega) \right\} d\omega + O_p\left(\frac{x_n}{n}\right),
\end{aligned}
$$

where

$$T_n = \frac{1}{2\pi} \int_{-\pi}^{\pi} \frac{I_{n,X}(\omega)}{\psi^2 \tilde{g}(\omega)} d\omega.$$

By Proposition 3.5 of Klüppelberg and Mikosch (1996) and Cramér–Wold device, we have

$$
\begin{pmatrix}
x_n \int_{-\pi}^{\pi} \frac{\partial f(\omega;\theta)}{\partial \theta_1}\big|_{\theta=\theta_0} \left\{ I_{n,X}(\omega) - T_n \psi^2 \tilde{g}(\omega) \right\} d\omega \\
\vdots \\
x_n \int_{-\pi}^{\pi} \frac{\partial f(\omega;\theta)}{\partial \theta_q}\big|_{\theta=\theta_0} \left\{ I_{n,X}(\omega) - T_n \psi^2 \tilde{g}(\omega) \right\} d\omega
\end{pmatrix}
$$

$$
\xrightarrow{\mathscr{L}}
\begin{pmatrix}
2 \sum_{t=1}^{\infty} S_t \left\{ \int_{-\pi}^{\pi} \frac{\partial f(\omega;\theta)}{\partial \theta_1}\big|_{\theta=\theta_0} \psi^2 \tilde{g}(\omega) \cos(t\omega) d\omega \right\} \\
\vdots \\
2 \sum_{t=1}^{\infty} S_t \left\{ \int_{-\pi}^{\pi} \frac{\partial f(\omega;\theta)}{\partial \theta_q}\big|_{\theta=\theta_0} \psi^2 \tilde{g}(\omega) \cos(t\omega) d\omega \right\}
\end{pmatrix}.
$$

Therefore, as $n \to \infty$, it holds that

$$x_n \boldsymbol{P}_n(\theta_0) \xrightarrow{\mathscr{L}} \boldsymbol{V}, \qquad (4.40)$$

where $\boldsymbol{V}$ is defined in Theorem 4.6. Thus, $\boldsymbol{P}_n(\theta_0) = O_p(x_n^{-1})$.

Now let us evaluate the orders of each term in (4.38) as follows:

$$O_p(\|\phi\|)\left[O_p(1) - O_p\left\{(\log n)^{2-2/\alpha}\right\} \cdot O_p(x_n^{-1})\right] \le O_p(x_n^{-1}). \qquad (4.41)$$

Note that, as $n \to \infty$, the following holds:

$$(\log n)^{2-2/\alpha} x_n^{-1} = (\log n)^{2-2/\alpha} \left(\frac{\log n}{n} \right)^{1/\alpha}$$

$$= \frac{1}{(\log n)^{1/\alpha}} \frac{(\log n)^2}{n^{1/\alpha}}$$

$$\to 0.$$

Therefore, from (4.41), we see that

$$O_p(\|\phi\|) \le O_p(x_n^{-1}). \tag{4.42}$$

Next, we evaluate the term $\boldsymbol{\varepsilon}$.

$$\frac{1}{n} \sum_{t=1}^{n} \|m(\lambda_t; \theta_0)\|^3 = \frac{1}{n} \sum_{t=1}^{n} \|m(\lambda_t; \theta_0)\| \|m(\lambda_t; \theta_0)\|^2$$

$$\le \frac{1}{n} \sum_{t=1}^{n} M_n \, m(\lambda_t; \theta_0)^{\mathsf{T}} m(\lambda_t; \theta_0)$$

$$= M_n \operatorname{tr} \{S_n(\theta_0)\}$$

$$= O_p \left\{ (\log n)^{2-2/\alpha} \right\}. \tag{4.43}$$

From (4.42) and (4.43), $\boldsymbol{\varepsilon}$ in (4.37) can be evaluated as

$$\|\boldsymbol{\varepsilon}\| \le \frac{1}{n} \sum_{t=1}^{n} \|m(\lambda_t; \theta)\|^3 \|\phi\|^2 |1 + Y_t|^{-1}.$$

Thus, we have

$$O_p(\|x_n \boldsymbol{\varepsilon}\|) = O_p \left\{ \frac{(\log n)^{2-1/\alpha}}{n^{1/\alpha}} \right\} = o_p(1). \tag{4.44}$$

Last, we consider the asymptotics of the empirical likelihood ratio statistic $R(\theta_0)$. Remember that

$$R(\theta) = \max_{w_1,\dots,w_n} \left\{ \prod_{t=1}^{n} n w_t ; \sum_{t=1}^{n} w_t \, m(\lambda_t; \theta) = 0, \ \sum_{t=1}^{n} w_t = 1, \ 0 \le w_1, w_2, \dots, w_n \le 1 \right\}.$$

Under $H: \theta = \theta_0$, $-2(x_n^2/n) \log R(\theta_0)$ can be expanded as

$$-2\frac{x_n^2}{n}\log R(\theta_0) = -2\frac{x_n^2}{n}\sum_{t=1}^{n}\log nw_t$$

$$= 2\frac{x_n^2}{n}\sum_{t=1}^{n}\log(1 + Y_t)$$

$$= 2\frac{x_n^2}{n}\sum_{t=1}^{n}Y_t - \frac{x_n^2}{n}\sum_{t=1}^{n}Y_t^2 + 2\frac{x_n^2}{n}\sum_{t=1}^{n}O_p(Y_t^3).$$

The first term can further be expanded as

$$2\frac{x_n^2}{n}\sum_{t=1}^{n}Y_t = 2\frac{x_n^2}{n}\sum_{t=1}^{n}\phi^{\mathrm{T}}m(\lambda_t; \theta_0)$$

$$= 2\frac{x_n^2}{n}\left\{S_n(\theta_0)^{-1}P_n(\theta_0) + \varepsilon\right\}^{\mathrm{T}}\sum_{t=1}^{n}m(\lambda_t; \theta_0)$$

$$= 2x_n^2\left\{P_n(\theta_0)^{\mathrm{T}}S_n(\theta_0)^{-1} + \varepsilon^{\mathrm{T}}\right\}P_n(\theta_0)$$

$$= 2\{x_nP_n(\theta_0)\}^{\mathrm{T}}S_n(\theta_0)^{-1}\{x_nP_n(\theta_0)\} + 2(x_n\varepsilon)^{\mathrm{T}}\{x_nP_n(\theta_0)\}.$$

The second term can further be expanded as

$$\frac{x_n^2}{n}\sum_{t=1}^{n}Y_t^2 = \frac{x_n^2}{n}\sum_{t=1}^{n}\left\{\phi^{\mathrm{T}}m(\lambda_t; \theta_0)\right\}^2$$

$$= x_n^2\phi^{\mathrm{T}}S_n(\theta_0)\phi$$

$$= x_n^2\left\{P_n(\theta_0)^{\mathrm{T}}S_n(\theta_0)^{-1} + \varepsilon^{\mathrm{T}}\right\}S_n(\theta_0)\left\{S_n(\theta_0)^{-1}P_n(\theta_0) + \varepsilon\right\}$$

$$= \{x_nP_n(\theta_0)\}^{\mathrm{T}}S_n(\theta_0)^{-1}\{x_nP_n(\theta_0)\}$$

$$\quad + (x_n\varepsilon)^{\mathrm{T}}S_n(\theta_0)(x_n\varepsilon) + 2(x_n\varepsilon)^{\mathrm{T}}\{x_nP_n(\theta_0)\}.$$

The last term can be evaluated as

$$\frac{x_n^2}{n}\left|\sum_{t=1}^{n}O_p(Y_t^3)\right| \le \frac{x_n^2}{n}{}^{\exists}c\sum_{t=1}^{n}|Y_t|^3$$

$$= \frac{x_n^2}{n}c\|\phi\|_E^3\sum_{t=1}^{n}\|m(\lambda_t; \theta_0)\|_E^3$$

$$= \frac{x_n^2}{n}O_p(x_n^{-3}) \cdot O_p\{n(\log n)^{2-2/\alpha})\}$$

$$= O_p\left\{\frac{(\log n)^{2-1/\alpha}}{n^{1/\alpha}}\right\}$$

$$= o_p(1).$$

Hence, noting (4.44), by (4.40) and Lemma 4.3, we have

$$-\frac{2x_n^2}{n} \log R(\theta_0) = \{x_n \boldsymbol{P}_n(\theta_0)\}^{\mathrm{T}} \boldsymbol{S}_n(\theta_0)^{-1} \{x_n \boldsymbol{P}_n(\theta_0)\} + o_p(1)$$

$$\xrightarrow{\mathscr{L}} \boldsymbol{V}^{\mathrm{T}} \boldsymbol{W}^{-1} \boldsymbol{V},$$

which completes the proof of Theorem 4.6. □

To apply the result in Theorem 4.6 in practice, we have to pay attention to the following remarks.

Remark 4.1 The limiting distribution (4.15) depends on the characteristic exponent α of the sαs random variables and unknown normalized power transfer function $\tilde{g}(\omega)$.

For the characteristic exponent α, we can construct an appropriate consistent estimator for it. For example, it is shown that Hill's estimator

$$\hat{\alpha}_{\mathrm{Hill}} = \left\{ \frac{1}{k} \sum_{t=1}^{k} \log \frac{|X|_{(t)}}{|X|_{(k+1)}} \right\}^{-1}$$

is a consistent estimator of α, where $|X|_{(1)} > \cdots > |X|_{(n)}$ is the order statistic of $|X(1)|, \cdots, |X(n)|$ and $k = k(n)$ is an integer satisfying some regularity conditions. (e.g., see Resnick and Stărică (1996) and Resnick and Stărică (1998).)

For the unknown normalized power transfer function $\tilde{g}(\omega)$, one can estimate it by the smoothed self-normalized periodogram $\tilde{I}_{n,X}$ with an appropriate weighting function $W_n(\cdot)$. It is weekly consistent to the normalized power transfer function. That is, for any $\omega \in [-\pi, \pi]$,

$$\tilde{J}_{n,X}(\omega) = \sum_{|k| \leq m} W_n(k) \tilde{I}_{n,X}(\lambda_k) \xrightarrow{\mathscr{P}} \tilde{g}(\omega), \quad \lambda_k = \omega + \frac{k}{n}, \quad |k| \leq m,$$

where the integer $m = m(n)$ satisfies $m \to \infty$ and $m/n \to 0$ as $n \to \infty$. (See Klüppelberg and Mikosch (1993), Theorem 4.1.) One possible choice of the weighting function $W_n(\cdot)$ and $m = m(n)$ are $W_n(k) = (2m+1)^{-1}$ and $m = [\sqrt{n}]$ ([x] denotes the integer part of x). We use this weighting function in Sect. 4.4 for numerical studies.

Remark 4.2 The confidence region for θ can be constructed in the similar way as described in Sect. 4.2. At the significant level q, one can define the region as

$$C_{q,n} = \left\{ \theta \in \Theta \; ; \; -\frac{2x_n^2}{n} \log R(\theta) < c_q \right\}, \tag{4.45}$$

where c_q is the $1 - q$ percentage point of the distribution of $\boldsymbol{V}^{\mathrm{T}} \boldsymbol{W} \boldsymbol{V}$, which can be obtained numerically.

4.4   Numerical Studies

In this section, we give the numerical simulation results for Theorem 4.6. Suppose that the observations $X(1), \cdots, X(n)$ are generated from the following scalar-valued stable MA(100) model:

$$X(t) = \sum_{j=0}^{100} \psi_j Z(t - j), \tag{4.46}$$

where $\{Z(t) : t \in \mathbb{Z}\}$ is a sequence of i.i.d. sαs random variables with scale $\sigma = 1$ and coefficients $\{\psi_j : j \in \mathbb{N}\}$ are defined as

$$\psi_j = \begin{cases} 1 & \text{for } j = 0, \\ b^j / j & \text{for } 1 \le j \le 100, \\ 0 & \text{otherwise.} \end{cases}$$

The process (4.46) cannot be expressed as the AR model or the ARMA models by finite coefficients without the trivial case. It is suitable to apply the empirical likelihood approach to making the statistical inference.

As described in Example 4.1, let us consider the estimation of the autocorrelation at lag 2 by the sample autocorrelation (SAC) method, i.e., from Davis and Resnick (1986), we have

$$\hat{\rho}(2) = \frac{\sum_{t=1}^{n-2} X(t)X(t + 2)}{\sum_{t=1}^{n} X(t)^2} \xrightarrow{\mathscr{P}} \rho(2). \tag{4.47}$$

The asymptotic distribution of $\hat{\rho}(2)$ in (4.47) can be found from the following general result. From Theorem 12.5.1 of Brockwell and Davis (1991), for fixed $l \in \mathbb{N}$,

$$x_n\{\hat{\rho}(l) - \rho(l)\} \xrightarrow{\mathscr{L}} \frac{\widetilde{S}_1}{\widetilde{S}_0} \left\{ \sum_{j=1}^{\infty} |\rho(l + j) + \rho(l - j) - 2\rho(j)\rho(l)|^\alpha \right\}^{1/\alpha},$$

where $\hat{\rho}(l) = \sum_{t=1}^{n-l} X(t)X(t + l) / \sum_{t=1}^{n} X(t)^2$, $\widetilde{S}_0$ and $\widetilde{S}_1$ are $\alpha/2$ and α-stable random variables, respectively.

On the other hand, the normalized power transfer function of the process (4.46) is given as

$$\tilde{g}(\omega) = \frac{\left| \sum_{j=0}^{100} \psi_j \exp(ij\omega) \right|^2}{\sum_{j=0}^{100} \psi_j^2}.$$

If we set the model $f(\omega; \theta)$ as $f(\omega; \theta) = |1 - \theta \exp(2i\omega)|^{-2}$, then we obtain

Table 4.1 90% confidence intervals (and length) for the autocorrelation with lag 2. Sample size is 300 and $\alpha = 1.5$. $b = 0.5$ in case 1 and $b = 0.9$ in case 2

	$\theta_0 \approx$	E.L			SAC		
Case 1.	0.1168	−0.0761	0.1930	(0.2691)	−0.0676	0.2481	(0.3157)
Case 2.	0.3603	0.1320	0.4765	(0.3445)	0.1388	0.5304	(0.3916)

Table 4.2 90% confidence intervals (and length) for the autocorrelation with lag 2. Sample size is 300 and $b = 0.5$

	α	E.L			SAC		
Case 3.	1.0	−0.1583	0.3335	(0.4918)	−0.1342	0.3891	(0.5233)
Case 4.	1.5	−0.0761	0.1930	(0.2691)	−0.0676	0.2481	(0.3157)
Case 5.	1.9	−0.0465	0.1329	(0.1794)	−0.0450	0.1365	(0.1815)

Table 4.3 90% confidence intervals (and length) for the autocorrelation with lag 2. $b = 0.5$, $\alpha = 1.5$, and $\theta_0 \approx 0.1168$

	n	E.L			SAC		
Case 6.	50	−0.2397	0.4313	(0.6710)	−0.2477	0.5629	(0.8106)
Case 7.	100	−0.3125	0.2228	(0.5353)	−0.3476	0.2218	(0.5694)

$$\theta_0 = \frac{\sum_{j=0}^{100} \psi_j \psi_{j+2}}{\sum_{j=0}^{100} \psi_j^2} = \rho(2). \tag{4.48}$$

Thus, it is natural to define the estimating function $m(\lambda_t; \theta)$ with this $f(\omega; \theta)$ for the empirical likelihood for $\rho(2)$.

In the numerical study, we construct a 90% confidence interval for θ_0 in (4.48) by the method of (4.45). The characteristic exponent α of sαs random variables is supposed to be known. We use the consistent estimator $\tilde{J}_{n,X}(\omega)$ with a weighting function W_n as described in Sect. 4.3 to estimate $\tilde{g}(\omega)$, and the cut-off point c_{10}, the 90 percentage point of the distribution $V^{\mathsf{T}}WV$, is computed via the Monte Carlo simulation for 10^5 times. Under this setting, we construct confidence intervals of $\theta_0 = \rho(2)$ from both the SAC method and the empirical likelihood method, and compare confidence intervals constructed by them. The numerical results are given in the following cases.

(i) $n = 300$, $\alpha = 1.5$, $b = 0.5$ in case 1 and $b = 0.9$ in case 2. From Table 4.1, we see that the length of the confidence intervals obtained by the empirical likelihood method is shorter than that by the SAC method.

(ii) $n = 300$ and $b = 0.5$. $\alpha = 1.0$ (Cauchy) in case 3, 1.5 in case 4 and 1.9 (near Gaussian) in case 5. From Table 4.2, we see that the larger α is, the better performance both methods show. In particular, the empirical likelihood method makes better inferences for θ_0 than the SAC method does in all cases.

Table 4.4 Coverage errors of
confidence intervals for the
parameter θ_0 in (4.48)

	Coverage errors	
	E.L.	SAC
Case 1.	0.082	0.087
Case 2.	0.089	0.096
Case 3.	0.094	0.098
Case 4.	0.082	0.087
Case 5.	0.053	0.056
Case 6.	0.092	0.095
Case 7.	0.086	0.090

(iii) $b = 0.5$ and $\alpha = 1.5$. $n = 50$ in case 6 and 100 in case 7. We consider the small
sample cases in the setting (iii). From Table 4.3, we see that even the sample
size n is small, the empirical likelihood method still works better than that the
SAC method in all cases.

At the end of Chap. 4, we focus on the coverage error to evaluate the performances
of the above confidence intervals. Let θ^U and θ^L be the endpoints of a confidence
interval. The coverage error is defined as

$$|\Pr[\{\theta_0 < \theta^L\} \cup \{\theta^U < \theta_0\}] - 0.1|. \tag{4.49}$$

We constructed the confidence intervals constructed for θ_0 by the Monte Carlo simulations for 1000 times. The coverage error (4.49) is evaluated by the empirical one.
Let (θ_l^L, θ_l^U) be the confidence intervals for $l = 1, \cdots, 1000$ in our simulations. For
each case above, we compute the following empirical coverage error:

$$\left| \frac{\sum_{l=1}^{1000} \mathbb{1}\{\theta_0 \notin (\theta_l^L, \theta_l^U)\}}{1000} - 0.1 \right|,$$

where $\mathbb{1}$ denotes the indicator function.

From Table 4.4, we see that the confidence intervals constructed by the empirical
likelihood are more accurate than those by the SAC method. Especially, it seems that
both methods give the coverage probabilities close to the nominal level when α is
near 2 (case 5). In addition, as n increases (from case 6 and case 7 to case 1), we can
see that the coverage error is decreasing.

Chapter 5
Self-weighted GEL Methods for Infinite Variance Processes

Abstract This chapter focuses on an alternative robust estimation/testing procedure for *possibly* infinite variance time series models. In the context of inference for
heavy-tailed observation, least absolute deviations (LAD) estimators are known to be
less sensitive to outliers than the classical least squares regression. This section gen-
eralizes the LAD regression-based inference procedure to the self-weighted version,
which is a concept originally introduced by Ling (2005) for AR processes. Using
the self-weighting method, we extend the generalized empirical likelihood (GEL)
method to possibly infinite variance process, and construct feasible and robust estima-
tion/testing procedures. The former half of this chapter provides a brief introduction
to the LAD regression method for possibly infinite variance ARMA models, and
construct the self-weighted GEL statistic following Akashi (2017). The desirable
asymptotic properties of the proposed statistics will be elucidated. The latter half of
this chapter illustrates an important application of the self-weighted GEL method to
the change point problem of time series models.

5.1 Introduction to Self-weighted Least Absolute Deviations Approach

In this section, we review the self-weighted least absolute deviations (LAD) method
by Ling (2005). Let us consider the following stationary ARMA (p, q) model:

$$X(t) = \sum_{j=1}^{p} b_j X(t - j) + \sum_{j=0}^{q} a_j \varepsilon(t - j), \qquad (5.1)$$

where $\{\varepsilon(t) : t \in \mathbb{Z}\}$ is a sequence of i.i.d. random variables and $a_0 = 1$. Here we
assume med$\{\varepsilon(t)\} = 0$ but do not assume any moment conditions at this stage. Now,
let us denote $b = (b_1, ..., b_p)^{\mathrm{T}}$, $a = (a_1, ..., a_q)^{\mathrm{T}}$ for general $b_j, a_j \in \mathbb{R}$ and set $d = p + q$. We also define $\theta_0 = (b_{01}, ..., b_{0p}, a_{01}, ..., a_{0q})^{\mathrm{T}}$, where b_{0j} $(j = 1, ..., p)$ and
a_{0j} $(j = 1, ..., q)$ are the true values of coefficients b_j and a_j. To guarantee the
stationarity of the model (5.1), we impose the following regularity conditions.

© The Author(s), under exclusive licence to Springer Nature Singapore Pte Ltd. 2018
Y. Liu et al., *Empirical Likelihood and Quantile Methods for Time Series*,
JSS Research Series in Statistics, https://doi.org/10.1007/978-981-10-0152-9_5

Assumption 5.1 (i) The characteristic polynomials $\phi(z; b) = 1 - b_1 z - \cdots - b_p z^p$ and $\psi(z; a) = 1 + a_1 z + \cdots + a_q z^q$ have no common zeros, and all roots of $\phi(z; b)$ and $\psi(z; a)$ are outside the unit circle for all $(b^T, a^T)^T \in \Theta$, where Θ is a compact subset of $\mathbb{R}^d$.

(ii) $\theta_0 \in \text{Int}(\Theta)$.

(iii) The innovations $\{\varepsilon(t) : t \in \mathbb{Z}\}$ have zero median and a bounded density f with $E[|\varepsilon(t)|^\tau] < \infty$ for some $\tau > 0$.

The problem concerned in this section is estimation problem of θ_0 and testing problem for a linear hypothesis of the form

$$H : R\theta_0 = c, \tag{5.2}$$

where R and c are, respectively, an $r \times d$ matrix and an $r \times 1$ vector ($r \leq d$). By choosing appropriate R and c, the hypothesis (5.2) includes various important problems. We introduce two examples as follows.

Example 5.1 (Test for serial correlation) If we choose $R = [O_{q \times p}, I_q]$ and $c = 0_q$, then (5.2) is equivalent to $H : a_1 = \cdots a_q = 0$. Therefore, we can test the model has nonzero-serial correlation or not.

Example 5.2 (Variable selection) Suppose that we want to test $H : b_j = 0$ for all $j \in \{j(1), ..., j(r)\}$, where $\{j(1), ..., j(r)\}$ is a subset of $\{1, ..., p\}$. This problem is captured by our framework (5.2) if we set $R = [K, O_{r \times q}]$ and $c = 0_r$, where $K = [u_{j(1)}, ..., u_{j(r)}]^T$ and u_j is the jth unit vector of $\mathbb{R}^p$.

As mentioned in the previous sections, if the model has heavy-tailed error term, the limit distributions of estimator and test statistics become intractable. As a result, it is often infeasible to determine cut-off points for confidence intervals or critical values for test statistics. To make classical statistics be robust, Ling (2005) considered, for the AR(p) model, the self-weighted LAD estimator for θ_0, which minimizes

$$\sum_{t=p+1}^{n} w_{t-1} \left| X(t) - \theta^T X_{t-1} \right|,$$

where $X_{t-1} = (X(t-1), ..., X(t-p))^T$. The weights $w_{t-1}, t = p+1, ..., n$ are of the form $w_{t-1} = w(X(t-1), ..., X(t-p))$, where $w : \mathbb{R}^p \to \mathbb{R}$ is a measurable function of X_{t-1}. The central roll of the self-weights is to control the leverage points $X(t)$ produced by $\varepsilon(t)$. Under some mild moment conditions of $\varepsilon(t)$, Ling (2005) showed asymptotic normality of the self-weighted LAD estimator for AR(p) models with possibly infinite variance. Motivated by the notion of self-weighting, Pan et al. (2007) extended Ling (2005)'s model to the ARMA model. Let us define

$$\varepsilon(t;\theta) = \begin{cases} 0 & (t \le 0) \\ \tilde{X}(t) - \sum_{j=1}^{p} b_j \tilde{X}(t-j) - \sum_{j=1}^{q} a_j \varepsilon(t-j;\theta) & (1 \le t \le n) \end{cases}, \quad (5.3)$$

$$\tilde{X}(t) = \begin{cases} 0 & (t \le 0) \\ X(t) & (t \ge 1) \end{cases}. \quad (5.4)$$

By the truncations (5.3) and (5.4), note that $\varepsilon(t;\theta_0) \ne \varepsilon(t)$ a.e. Pan et al. (2007) proposed the self-weighted LAD estimator for ARMA model (5.1) as

$$\hat{\theta}_{\mathrm{LAD}} := \arg\min_{\theta \in \Theta} \sum_{t=u+1}^{n} \tilde{w}_{t-1} |\varepsilon(t;\theta)|, \quad (5.5)$$

where $u \ge \max\{p, q\} + 1$ and $\tilde{w}_{t-1}$ is the self-weight for ARMA model defined as

$$\tilde{w}_{t-1} = \left(1 + \sum_{k=1}^{t-1} k^{-\gamma} |X(t-k)|\right)^{-2} \quad (5.6)$$

with $\gamma > \max\{2, 2/\tau\}$ and the same τ as in Assumption 5.1.[1] Under Assumption 5.1 and some regularity conditions for f, Pan et al. (2007) showed that the self-weighted LAD estimator (5.5) has asymptotic normality, and the Wald test statistic based on $\hat{\theta}_{\mathrm{LAD}}$ converges to a standard χ^2 distribution. Since the limit distributions of the statistics proposed by Ling (2005) and Pan et al. (2007) are pivotal, it is easy to detect the critical values for the test statistics even though the thickness of the innovation term is unknown.

5.2 Self-weighted GEL Statistics

As seen in the precious subsection, the self-weighting method proposed by Ling (2005) enables us to do the hypothesis test for possibly infinite variance processes. However, the Wald test statistics proposed in Ling (2005) and Pan et al. (2007) contain kernel-type estimators of the density function of the innovation process, and the choice of the bandwidth parameter is not clear. To avoid such involved problem, this section introduces the self-weighted GEL estimator for θ_0 and a test statistic for the hypothesis (5.2). To construct the robust GEL method for ARMA model (5.1), let us define the followings:

---

[1] Pan et al. (2007) proposed another self-weight of more general form $\tilde{w}_{t-1} = (1 + \sum_{k=1}^{t-1} k^{-\gamma} (\log k)^{\delta} |X(t-k)|)^{-\alpha}$ $(\gamma > 2, \alpha \ge 2, \delta \ge 0)$. However, the optimal choice of the parameters in self-weights is highly nontrivial, so we confine ourselves to the self-weight of the form (5.6) to keep the focus of this section.

$$A_{t-1}(\theta) = \frac{\partial \varepsilon(t; \theta)}{\partial \theta} \quad \text{and} \quad B_{t-1}(\theta) = \begin{pmatrix} A_{t-1}(\theta) \\ \varphi_{t-1} \end{pmatrix},$$

where φ_{t-1} is an s-dimensional measurable function of $\{X(i) : i \leq t - 1\}$. We can choose the function φ_{t-1} arbitrarily provided that the following condition is satisfied.

Assumption 5.2 $\sup_{t=u+1,\ldots,n} \|\tilde{w}_{t-1}\varphi_{t-1}\| \leq c$ for some $c > 0$.

Motivated by the definition of the LAD estimator (5.5), we consider the subgradient of the object function in the right-hand side of (5.5), i.e.,

$$g_t^*(\theta) := \tilde{w}_{t-1}\text{sign}\{\varepsilon(t; \theta)\}B_{t-1}(\theta). \tag{5.7}$$

To see the motivation of (5.7), let us consider the purely AR(p) model and set $s = 0$ (we call the moment function (5.7) with $s = 0$ and $s > 0$, respectively, the *just-* and *over*-identified moment function). By a simple calculation, we get $\varepsilon(t; \theta) = X(t) - \theta^{\mathrm{T}}X_{t-1}$ for $t \geq u + 1$ and

$$E[g_t^*(\theta)] = -2E\left[\left\{F((\theta - \theta_0)^{\mathrm{T}}X_{t-1}) - \frac{1}{2}\right\}\tilde{w}_{t-1}X_{t-1}\right], \tag{5.8}$$

where F is the distribution function of $\varepsilon(t)$. Since we assume zero median of $\varepsilon(t)$ ($F(0) = 1/2$), (5.8) is equal to zero at $\theta = \theta_0$. Therefore, we can construct the empirical likelihood ratio statistic based on the moment function (5.7) as

$$R_n(\theta) = \frac{\sup\left\{\prod_{t=u+1}^n v_t : \sum_{t=u+1}^n v_t = 1, \ \sum_{t=u+1}^n v_t g_t^*(\theta) = 0_m\right\}}{\{1/(n-u)\}^{n-u}},$$

where we define $m = p + q + s$. By the Lagrange multiplier method, we get the log-empirical likelihood function at θ as

$$-2\log R_n(\theta) = 2 \sup_{\lambda \in \Lambda_{\text{EL}}(\theta)} \sum_{t=u+1}^n \log\left\{1 - \lambda^{\mathrm{T}}g_t^*(\theta)\right\}, \tag{5.9}$$

where $\Lambda_{\text{EL}}(\theta) = \{\lambda \in \mathbb{R}^m : \lambda^{\mathrm{T}}g_t^*(\theta) < 1, \ t = u + 1, \ldots, n\}$. Motivated by the representation (5.9), the self-weighted GEL test statistic for the linear hypothesis (5.2) is naturally defined as

$$r_{\rho,n}^* = 2\left[\inf_{R\theta=c} \sup_{\lambda \in \hat{\Lambda}_n(\theta)} P_n^*(\theta, \lambda) - \inf_{\theta \in \Theta} \sup_{\lambda \in \hat{\Lambda}_n(\theta)} P_n^*(\theta, \lambda)\right], \tag{5.10}$$

where

$$P_n^*(\theta, \lambda) = \sum_{t=u+1}^n \rho\left\{\lambda^{\mathrm{T}}g_t^*(\theta)\right\}$$

for a certain continuous and concave function ρ on its domain $\mathcal{V}_\rho$ and

$$\hat{\Lambda}_n(\theta) = \left\{ \lambda \in \mathbb{R}^m : \lambda^T g_t^*(\theta) \in \mathcal{V}_\rho, \ t = u + 1, \dots, n \right\}.$$

We also consider the maximum GEL estimators for θ_0 and the Lagrange multiplier as

$$\hat{\theta}^* = \arg\min_{\theta \in \Theta} \sup_{\lambda \in \hat{\Lambda}_n(\theta)} P_n^*(\theta, \lambda) \quad \text{and} \quad \hat{\lambda}^* = \arg\max_{\lambda \in \hat{\Lambda}_n(\hat{\theta}^*)} P_n^*(\hat{\theta}^*, \lambda),$$

respectively.

Remark 5.1 Note that the proposed test statistic $r_{\rho,n}^*$ contains important statistics in econometrics such as the empirical likelihood (Owen 1988), the continuous updating GMM (Hansen et al. 1996) and the exponential tilting method (Kitamura and Stutzer 1997). In particular, the above three methods are included in the Cressie–Read power divergence family (Cressie and Read 1984), which is defined as

$$\rho_{CR}(v) = -\frac{1}{c+1} \left\{ (1 + cv)^{(c+1)/c} - 1 \right\} = \begin{cases} \log(1 - v) & (c \to -1) \\ 1 - \exp(v) & (c \to 0) \\ -(v + 2)v/2 & (c = 1) \end{cases}.$$

Generally, we can choose ρ arbitrarily according to the following condition.

Assumption 5.3 ρ is twice continuously differentiable, concave on $\mathcal{V}_\rho$, and satisfies $\rho(0) = 0$ and $\rho'(0) = \rho''(0) = -1$.

To derive the limit distributions of $\hat{\theta}^*$, $\hat{\lambda}^*$ and $r_{\rho,n}^*$, we need some additional conditions.

Assumption 5.4 (i) θ_0 is a unique solution to $E[g_t^{*0}(\theta)] = 0_m$, where $g_t^{*0}(\theta) = \delta_{t-1}\text{sign}\{\varepsilon(t, \theta)\}B_{t-1}(\theta)$ and $\delta_{t-1} = (1 + \sum_{k=1}^\infty k^{-\gamma}|X(t - k)|)^{-2}$.
(ii) $\Omega = E[\delta_{t-1}^2 Q_{t-1} Q_{t-1}^T]$ is nonsingular and $G = G(\theta_0)$ is of full-column rank. Here

$$G(\theta) = \frac{\partial E[g_t^{*0}(\theta)]}{\partial \theta^T},$$

$Q_{t-1} = (U_{t-1}, \dots, U_{t-p}, V_{t-1}, \dots, V_{t-q}, \varphi_{t-1}^T)^T$, $\{U_t : t \in \mathbb{Z}\}$ and $\{V_t : t \in \mathbb{Z}\}$ are autoregressive processes defined as $\phi(B; b_0)U_t = \varepsilon(t)$ and $\psi(B; a_0)V_t = \varepsilon(t)$.

The condition (i) in Assumption 5.4 guarantees consistency of the GEL estimator. The latter condition requires the self-weights to satisfy additional moment condition, and is used to control the stochastic order of some remainder terms. Under the assumptions, we get the following result.

Theorem 5.5 *Suppose that Assumptions 5.1–5.4 hold.*

(i) $n^{1/2}((\hat{\theta}^* - \theta_0)^T, \hat{\lambda}^{*T})^T$ is asymptotically normal with mean 0_{d+m} and variance

$$\begin{bmatrix} \Omega_\theta & O_{d \times m} \\ O_{m \times d} & \Omega_\lambda \end{bmatrix},$$

where $\Omega_\theta = (G^T \Omega^{-1} G)^{-1}$ and $\Omega_\lambda = (I_m - \Omega^{-1} G (G^T \Omega^{-1} G)^{-1} G^T) \Omega^{-1}$.

(ii) Under $H : R\theta_0 = c$, $r^*_{\rho,n} \to \chi^2_r$ in law as $n \to \infty$.

Remark 5.2 Regardless of whether the model has finite variance or not, the limit distribution in (ii) of the GEL test statistic is invariant, i.e., the test based on $r^*_{\rho,n}$ is asymptotically independent of the thickness of the error distribution or the choice of the self-weights. As seen in the previous section, it is often difficult to detect tail behavior of real data correctly. Therefore, this robust feature of the self-weighted test statistic is desirable in practice.

Before the proof of Theorem 5.5, let us impose two important lemmas. The first one shows the uniform convergence of the sum of the moment functions. To show the consistency of the GEL estimators, it is often required to show the uniform convergence $\|\hat{g}^*_n(\theta) - \bar{g}^*_n(\theta)\| = o_p(1)$ uniformly in $\theta \in \Theta$, where

$$\hat{g}^*_n(\theta) = \frac{1}{n-u} \sum_{t=u+1}^{n} g^*_t(\theta) \quad \text{and} \quad \bar{g}^*_n(\theta) = E\left[\hat{g}^*_n(\theta)\right].$$

Since $\hat{g}^*_n(\theta)$ contains nonsmooth part with respect to θ, it is not easy to use Corollary 2.2 of Newey (1991). It is also hard to show the stochastic equicontinuity of $\{\hat{g}^*_n(\theta) : n \geq 1\}$ directly. Therefore, we need to generalize the methodology of Tauchen (1985), which is based on Huber (1967) to dependent case, and get the following lemma.

Lemma 5.1 $\sup_{\theta \in \Theta} \|\hat{g}^*_n(\theta) - \bar{g}^*_n(\theta)\| = o_p(1)$.

The second lemma shows that $P^*_n(\theta, \lambda)$ is well approximated by some smooth function near its optima. Let us define

$$L^*_n(\theta, \lambda) = -n\{G(\theta - \theta_0) + \hat{g}^*(\theta_0)\}^T \lambda - \frac{n}{2} \lambda^T \Omega \lambda.$$

Furthermore, hereafter we define

$$\hat{\theta}^L = \arg\min_{\theta \in \Theta} \sup_{\lambda \in \mathbb{R}^m} L^*_n(\theta, \lambda) \quad \text{and} \quad \hat{\lambda}^L = \arg\max_{\lambda \in \mathbb{R}^m} L^*_n(\tilde{\theta}, \lambda).$$

Lemma 5.2 $P^*_n(\hat{\theta}^*, \hat{\lambda}^*) = L^*_n(\hat{\theta}^L, \hat{\lambda}^L) + o_p(1)$.

The proofs of Lemmas 5.1 and 5.2 are relegated to the last section of this chapter. By using these lemmas, we get Theorem 5.5 as follows.

Proof of Theorem 5.5. Throughout this chapter, C will denote a generic positive constant that may be different in different uses and "with probability approaching one" will be abbreviated as w.p.a.1. First, we show the assertion (i) by establishing the relations $n^{1/2}(\hat{\theta}^* - \hat{\theta}^L) = o_p(1)$ and $n^{1/2}(\hat{\lambda}^* - \hat{\lambda}^L) = o_p(1)$. For $\hat{\theta}^*$, by $L_n^*(\hat{\theta}^*, \hat{\lambda}^*) - L_n^*(\hat{\theta}^L, \hat{\lambda}^*) = o_p(1)$ (see the statement (ii) in the proof of Lemma 5.2), we have

$$o_p\left(\frac{1}{n}\right) = \frac{1}{n}(L_n^*(\hat{\theta}^*, \hat{\lambda}^*) - L_n^*(\hat{\theta}^L, \hat{\lambda}^*)) = -\left(\hat{\theta}^* - \hat{\theta}^L\right)^{\mathrm{T}} G^{\mathrm{T}} \hat{\lambda}^*.$$

Since G is of full rank and $\hat{\lambda}^* = O_p(n^{-1/2})$ by Lemmas 5.6 and 5.8, we get $\hat{\theta}^* - \hat{\theta}^L = o_p(n^{-1/2})$. For $\hat{\lambda}^*$, from (iii) in the proof of Lemma 5.2, (5.42) and (5.43), we obtain

$$o_p\left(\frac{1}{n}\right) = \frac{1}{n}(L_n^*(\hat{\theta}^L, \hat{\lambda}^*) - L_n^*(\hat{\theta}^L, \hat{\lambda}^L)) = -\frac{1}{2}\left(\hat{\lambda}^* - \hat{\lambda}^L\right)^{\mathrm{T}} \Omega \left(\hat{\lambda}^* - \hat{\lambda}^L\right).$$

Since Ω is nonsingular, $\hat{\lambda}^* - \hat{\lambda}^L = o_p(n^{-1/2})$. As in the proof of Lemma 1 of Li et al. (2011) and by Assumption 5.2, we have

$$n^{1/2}\hat{g}^*(\theta_0) - \frac{1}{n^{1/2}} \sum_{t=u+1}^{n} g_t^* = o_p(1),$$

where $g_t^* = \delta_{t-1}\mathrm{sign}(\varepsilon(t))Q_{t-1}$. Since g_t^* is a stationary ergodic square-integrable sequence of martingale differences with respect to $\mathscr{F}_t = \sigma\{e_s : s \leq t\}$, we get $n^{-1/2} \sum_{t=u+1}^{n} g_t^* \to N(0_m, \Omega)$ in law by Theorem 3.2 of Hall and Heyde (1980). Therefore, we have $n^{1/2}\hat{g}_n^*(\theta_0) \to N(0_m, \Omega)$ in law and get the desired result from (5.45).

Second, we show the assertion (ii). By Lemma 5.2, we expand the second part of (5.10) as

$$2P_n^*(\hat{\theta}^*, \hat{\lambda}^*) = 2L_n^*(\hat{\theta}^L, \hat{\lambda}^L) + o_p(1)$$
$$= n\hat{g}^*(\theta_0)^{\mathrm{T}} \left(\Omega - \Omega^{-1}G\Sigma G^{\mathrm{T}}\Omega^{-1}\right) \hat{g}^*(\theta_0) + o_p(1)$$
$$= \left\{n^{1/2}\Omega^{-1/2}\hat{g}^*(\theta_0)\right\}^{\mathrm{T}} \Lambda \left\{n^{1/2}\Omega^{-1/2}\hat{g}^*(\theta_0)\right\} + o_p(1), \qquad (5.11)$$

where $\Lambda = I_m - \Omega^{-1/2}G\Sigma G^{\mathrm{T}}\Omega^{-1/2}$. On the other hand, by the same argument as above,

$$2 \inf_{R\theta=c} \sup_{\lambda \in \hat{\Lambda}_n(\theta)} P_n^*(\theta, \lambda) = 2L_n^*(\hat{\theta}^{L,r}, \hat{\lambda}^{L,r}) + o_p(1),$$

where $\hat{\theta}^{L,r} = \arg\min_{R\theta=c} \sup_{\lambda \in \mathbb{R}^m} \hat{L}_n(\theta, \lambda)$ and $\hat{\lambda}^{L,r} = \arg\max_{\lambda \in \mathbb{R}^m} \hat{L}_n(\hat{\theta}^{L,r}, \lambda)$. By the Lagrange multiplier method for $L_n^*(\hat{\theta}^{L,r}, \hat{\lambda}^{L,r})$, we get the expansion

$$2 \inf_{R\theta=c} \sup_{\lambda \in \hat{\Lambda}_n(\theta)} P_n^*(\theta, \lambda) = \left\{ n^{1/2} \Omega^{-1/2} \hat{g}^*(\theta_0) \right\}^{\mathrm{T}} \Lambda^r \left\{ n^{1/2} \Omega^{-1/2} \hat{g}^*(\theta_0) \right\} + o_p(1),$$

(5.12)

where $\Lambda^r = I_m - \Omega^{-1/2} G P^r G^{\mathrm{T}} \Omega^{-1/2}$ and $P^r = \Sigma - \Sigma R^{\mathrm{T}} (R \Sigma R^{\mathrm{T}})^{-1} R \Sigma$.
By (5.11) and (5.12), $r_{\rho,n}^*$ admits the expansion

$$r_{\rho,n}^* = \left\{ n^{1/2} \Omega^{-1/2} \hat{g}_n^*(\theta_0) \right\}^{\mathrm{T}} (\Lambda^r - \Lambda) \left\{ n^{1/2} \Omega^{-1/2} \hat{g}_n^*(\theta_0) \right\} + o_p(1).$$

It is also easily shown that $(\Lambda^r - \Lambda)^2 = \Lambda^r - \Lambda$ and rank$(\Lambda^r - \Lambda) = r$. Therefore, $r_{\rho,n}^* \to \chi_r^2$ in law as $n \to \infty$ by Rao and Mitra (1971).   □

5.3 Application to the Change Point Test

An important application of Theorem 5.5 is the change point detection problem of infinite variance ARMA models. Suppose that an observed stretch $\{X(1), ..., X(n)\}$ is generated from the process (5.1) with a time-varying coefficient

$$\theta_{t,n} := (b_{t,n,1}, ..., b_{t,n,p}, a_{t,n,1}, ..., a_{t,n,q})^{\mathrm{T}}.$$

To describe the change point problem, we assume that the coefficient vector $\theta_{t,n}$ satisfies

$$\theta_{t,n} = \begin{cases} \theta_1 & (t = 1, ..., t_0) \\ \theta_2 & (t = t_0 + 1, ..., n) \end{cases}$$

with some fixed vectors $\theta_1, \theta_2 \in \mathbb{R}^d$. Here t_0 is the unknown change point satisfying $t_0 = \lfloor u_0 n \rfloor$ with a fixed real number $u_0 \in (0, 1)$. Under this setting, the change point problem is described by the hypothesis testing

$$H : \theta_1 = \theta_2 = {}^\exists \theta_0 \quad \text{against} \quad A : \theta_1 \neq \theta_2. \tag{5.13}$$

To test (5.13), Akashi et al. (2018) defined the empirical likelihood statistic as

$$L_{n,k}(\theta_1, \theta_2) := \sup \left\{ \left(\prod_{i=1}^k v_i \right) \left(\prod_{j=k+1}^n v_j \right) : (v_1, ..., v_n) \in \mathscr{P}_{n,k} \cap \mathscr{M}_{n,k}(\theta_1, \theta_2) \right\},$$

where $\mathscr{P}_{n,k}$ and $\mathscr{M}_{n,k}(\theta_1, \theta_2)$ are subsets of the cube $[0, 1]^n$ defined as

$$\mathscr{P}_{n,k} := \left\{ (v_1, ..., v_n) \in [0, 1]^n : \sum_{i=1}^k v_i = \sum_{j=k+1}^n v_j = 1 \right\}$$

and

$$\mathcal{M}_{n,k}(\theta_1, \theta_2) := \left\{ (v_1, ..., v_n) \in [0, 1]^n : \sum_{i=1}^{k} v_i g_i^*(\theta_1) = \sum_{j=k+1}^{n} v_j g_j^*(\theta_2) = 0_m \right\}.$$

To simplify the problem and avoid complicity, we confine the moment function for the GEL statistics within the just-identified case, i.e.,

$$g_t^*(\theta) := \tilde{w}_{t-1} \text{sign}\{\varepsilon(t; \theta)\} A_{t-1}(\theta) = \tilde{w}_{t-1} \text{sign}\{\varepsilon(t; \theta)\} \frac{\partial \varepsilon(t; \theta)}{\partial \theta}$$

and hence $m = d = p + q$. Note that the unconstrained maximum empirical likelihood is represented as

$$L_{n,k,E} := \sup \left\{ \prod_{i=1}^{n} v_i : (v_1, ..., v_n) \in \mathscr{P}_{n,k} \right\} = k^{-k}(n-k)^{-(n-k)},$$

and hence, the logarithm of the empirical likelihood ratio statistic is given by

$$-\log \frac{L_{n,k}(\theta_1, \theta_2)}{L_{n,k,E}} \tag{5.14}$$

$$= -\log \sup \left\{ \left(\prod_{i=1}^{k} k v_i \right) \left(\prod_{j=k+1}^{n} (n-k) v_j \right) : (v_1, ..., v_n) \in \mathscr{P}_{n,k} \cap \mathcal{M}_{n,k}(\theta_1, \theta_2) \right\}$$

$$= \left[\sup_{\lambda} \sum_{t=1}^{k} \log \left\{ 1 - \lambda^{\mathrm{T}} g_t^*(\theta_1) \right\} + \sup_{\eta} \sum_{t=k+1}^{n} \log \left\{ 1 - \eta^{\mathrm{T}} g_t^*(\theta_2) \right\} \right]. \tag{5.15}$$

The former part of the quantity (5.15) is regarded as the negative empirical likelihood ratio for the observation $\{X(1), ..., X(k)\}$ at θ_1, while the later one is for $\{X(k+1), ..., X(n)\}$ at θ_2. Motivated by the concept of GEL, we define the GEL statistic at θ_1 and θ_2, respectively, before and after a plausible change point $t = k$, as

$$l_{n,k}^{\rho}(\theta_1, \theta_2) = \left[\sup_{\lambda} \sum_{t=1}^{k} \rho \left\{ \lambda^{\mathrm{T}} g_t^*(\theta_1) \right\} + \sup_{\eta} \sum_{t=k+1}^{n} \rho \left\{ \eta^{\mathrm{T}} g_t^*(\theta_2) \right\} \right]. \tag{5.16}$$

Based on the GEL statistic (5.16), we finally define the test statistic for the change point problem (5.13). Since the maximum GEL under H at a plausible change point $t = k$ is given by

$$P_{n,k}^{\rho} := \inf_{\theta \in \Theta} \{ l_{n,k}^{\rho}(\theta, \theta) \},$$

one may define the empirical likelihood ratio test statistic by

$$T_n^\rho := 2 \max_{\lfloor r_1 n \rfloor \le k \le \lfloor r_2 n \rfloor} P_{n,k}^\rho,$$

where $0 < r_1 < r_2 < 1$ for fixed constants. Note that we do not consider the maximum of $\{P_{n,k}^\rho : k = 1, \ldots, n\}$ as $P_{n,k}^\rho$ cannot be estimated accurately for small and large values of k.

Remark 5.3 Note that our definition of $P_{n,k}^\rho$ is different from that in some literature. The numerator in (5.14) represents the empirical likelihood under the assumption that "the plausible change does not occur at $t = k$." The denominator represents the empirical likelihood under the assumption that "the plausible change occurs at $t = k$." We do not put any restriction between the parameters before and after the change under the alternative. Thus, our approach differs from a part of the literature, which considers the statistic

$$\frac{\sup_\theta L_{n,k}(\theta, \theta)}{\sup_{\theta_1, \theta_2} L_{n,k}(\theta_1, \theta_2)} \tag{5.17}$$

(see Chuang and Chan 2002, or Ciuperca and Salloum 2015). In other words, we work with a more general alternative that yields substantial computational advantages because the calculation of the supremum in the denominator of (5.17) corresponds to a $2d$-dimensional optimization problem, which has to be solved for each k.

To show the convergence of the test statistic, we need an additional condition for the dependence structure of the model as follows.

Assumption 5.6 (i) There exists a constant $\iota > 2$ such that $E[\|g_t^*(\theta_0)\|^\iota] < \infty$.
(ii) The sequence $\{\delta_{t-1}\text{sign}(\varepsilon(t))Q_{t-1} : t \in \mathbb{Z}\}$ is strong mixing with mixing coefficients α_l that satisfy $\sum_{l=1}^\infty \alpha_l^{1-2/\iota} < \infty$.

In this case, the maximum GEL estimator $\hat{\theta}_{n,k}$ is defined by

$$\hat{\theta}_{n,k} = \arg\min_{\theta \in \Theta}\{l_{n,k}(\theta, \theta)\},$$

and the consistency with corresponding rate of convergence of this statistic is given in the following theorem.

Theorem 5.7 *Suppose that Assumptions 5.1–5.4 and 5.6 hold and define* $k^* := rn$ *for some* $r \in (0, 1)$. *Then, under the null hypothesis H, we have, as* $n \to \infty$, $\hat{\theta}_{n,k^*} - \theta_0 = O_p(n^{-1/2})$.

As seen from Theorem 5.7, T_n^ρ is not accurate for small k and $n - k$ as the result does not hold if $k/n = o(1)$ or $(n - k)/n = o(1)$. In addition, the empirical likelihood ratio statistic is not computable for small k and $n - k$. For this reason, hereafter, we consider the trimmed and weighted version of empirical likelihood ratio test statistic, defined by

$$\tilde{T}_n^\rho := 2 \max_{k_{1n} \le k \le k_{2n}} h\left(\frac{k}{n}\right) P_{n,k}^\rho, \tag{5.18}$$

where h is a given weight function, $k_{1n} := r_1 n$, $k_{2n} := r_2 n$ and $0 < r_1 < r_2 < 1$. If $\tilde{T}_n$ takes a significant large value, we have enough reason to reject the null hypothesis H of no change point. We also need a further assumption to control a remainder terms in the stochastic expansion of $\tilde{T}_n^\rho$.

Assumption 5.8 $\sup_{0<r<1} h(r)^2 < \infty$.

With this additional assumption, the limit distribution of the test statistic (5.18) can be derived in the following theorem.

Theorem 5.9 *Suppose that Assumptions 5.1–5.4, 5.6 and 5.8 hold.*

(i) Under the null hypothesis H of no change point,

$$\tilde{T}_n^\rho \to \tilde{T} := \sup_{r_1 \le r \le r_2} \left\{ r^{-1}(1-r)^{-1} h(r) \| B(r) - r B(1) \|^2 \right\} \tag{5.19}$$

in law as $n \to \infty$, where $\{B(r) : r \in [0, 1]\}$ is a d-dimensional vector of independent Brownian motions.

(ii) Under the alternative $A : \theta_1 \ne \theta_2$, we have $\tilde{T}_n \to \infty$ in probability as $n \to \infty$.

The proofs of Theorems 5.5, 5.7 and 5.9 share a lot of similar argument, so we omit the detail. The proof is obtained by a slight modification of Akashi et al. (2018). A test for the hypotheses in (5.13) is now easily obtained by rejecting the null hypothesis in (5.13) whenever

$$\tilde{T}_n^\rho > q_{1-\alpha}, \tag{5.20}$$

where $q_{1-\alpha}$ is the $(1-\alpha)$-quantile of the distribution of the random variable $\tilde{T}$ defined on the right-hand side of equation (5.19). The statement (ii) in Theorem 5.9 shows that the power of the test (5.20) approaches 1 at any fixed alternative. In other words, the test is consistent.

5.4 Numerical Studies

This section illustrates the finite sample performance of the GEL change point test (5.20), and compares the goodness of the proposed method with CUSUM-type test by Qu (2008) via simulation study. Suppose that an observed stretch $\{X(1), ..., X(n)\}$ is generated from the AR(1) model

$$X(t) = b_{t,n} X(t-1) + \varepsilon(t) \quad (t = 1, ..., n), \tag{5.21}$$

where the parameter $b_{t,n}$ satisfies

$$b_{t,n} = \begin{cases} b_1 & (t = 1, ..., \lfloor u_0 n \rfloor) \\ b_2 & (t = \lfloor u_0 n \rfloor + 1, ..., n) \end{cases},$$

and the unknown change point parameter is set as $u_0 = 0.5$ and 0.8. The coefficient b_1 is fixed as $b_1 = 0.3$, and we tried simulations with various values of b_2 shown in tables below. The innovation $\varepsilon(t)$ follows one of i.i.d. standard normal, t_2 and Cauchy distributions. Throughout this simulation, the nominal level is chosen as $\alpha = 0.05$. To calculate the self-weighted GEL statistics, we use the weight function proposed by Ling (2005), which is defined as

$$\tilde{w}_{t-1} = \begin{cases} 1 & (d_{t-1} = 0) \\ (c_0/d_{t-1})^3 & (d_{t-1} \neq 0) \end{cases},$$

where $d_{t-1} = |X(t-1)| \mathbb{I}(|X(t-1)| > c_0)$ and c_0 is the 95% quantile of $\{X(1), ..., X(n)\}$. Under the settings, we generate samples from the model (5.21) with length $n = 100, 200, 400$, and calculate the simulated rejection rate based on 1000 times iteration. The results are shown in Tables 5.1, 5.2 and 5.3. The simulated rejection rate based on CUSUM-type test by Qu (2008) is labeled as "SQ". The same tables also display the results by the GEL test (5.20) with uniform weight function $h(u) \equiv 1$.

From the simulation, we observe that the rejection rate of the SQ-test is slightly conservative (see the cases of $b_2 = 0.3$, i.e., the case that the model has no change point). On the other hand, the simulated type-I error rate of GEL test (5.18) approximates the true nominal level well overall, and the accuracy of the GEL test is improved as the sample size grows. Especially, the type-I error rate of the self-weighted GEL test is less sensitive to the thickness of the tail distribution of the error terms than SQ-test.

Table 5.1 Simulated rejection rate of SQ and $\tilde{T}_n^\rho$ test with normal-distributed innovations ($b_1 = 0.3$)

$u_0 = 0.5$		b_2	−0.3	0.0	0.3	0.6	0.9	$u_0 = 0.8$		b_2	−0.3	0.0	0.3	0.6	0.9
$n = 100$	SQ		0.422	0.090	0.024	0.069	0.203	$n = 100$	SQ		0.104	0.040	0.028	0.079	0.444
	$\tilde{T}_n^\rho$		0.327	0.087	0.040	0.084	0.248		$\tilde{T}_n^\rho$		0.140	0.058	0.050	0.134	0.597
$n = 200$	SQ		0.794	0.254	0.049	0.228	0.713	$n = 200$	SQ		0.291	0.081	0.049	0.183	0.847
	$\tilde{T}_n^\rho$		0.731	0.235	0.075	0.205	0.729		$\tilde{T}_n^\rho$		0.328	0.069	0.061	0.232	0.861
$n = 400$	SQ		0.991	0.507	0.044	0.565	0.994	$n = 400$	SQ		0.682	0.142	0.029	0.329	0.989
	$\tilde{T}_n^\rho$		0.973	0.433	0.053	0.467	0.993		$\tilde{T}_n^\rho$		0.754	0.168	0.030	0.393	0.984

Table 5.2 Simulated rejection rate of SQ and $\tilde{T}_n^\rho$ test with t_2-distributed innovations ($b_1 = 0.3$)

$u_0 = 0.5$	b_2	-0.3	0.0	0.3	0.6	0.9	$u_0 = 0.8$	b_2	-0.3	0.0	0.3	0.6	0.9
$n = 100$	SQ		0.572	0.138	0.028	0.158	0.409	$n = 100$	SQ		0.129	0.046	0.029	0.113	0.592
	$\tilde{T}_n^\rho$		0.480	0.147	0.049	0.186	0.498		$\tilde{T}_n^\rho$		0.202	0.059	0.050	0.228	0.795
$n = 200$	SQ		0.916	0.466	0.027	0.532	0.783	$n = 200$	SQ		0.357	0.080	0.029	0.278	0.912
	T_n		0.910	0.384	0.055	0.440	0.817		$\tilde{T}_n^\rho$		0.531	0.125	0.051	0.424	0.978
$n = 400$	SQ		0.985	0.847	0.026	0.911	0.950	$n = 400$	SQ		0.682	0.240	0.029	0.593	0.977
	$\tilde{T}_n^\rho$		0.997	0.719	0.056	0.842	0.988		$\tilde{T}_n^\rho$		0.909	0.317	0.054	0.709	0.998

Table 5.3 Simulated rejection rate of SQ, and $\tilde{T}_n^\rho$ test with Cauchy-distributed innovations ($b_1 = 0.3$)

$u_0 = 0.5$	b_2	-0.3	0.0	0.3	0.6	0.9	$u_0 = 0.8$	b_2	-0.3	0.0	0.3	0.6	0.9
$n = 100$	SQ		0.317	0.126	0.029	0.220	0.429	$n = 100$	SQ		0.069	0.043	0.023	0.171	0.537
	$\tilde{T}_n^\rho$		0.448	0.172	0.062	0.341	0.724		$\tilde{T}_n^\rho$		0.242	0.085	0.060	0.355	0.819
$n = 200$	SQ		0.539	0.300	0.029	0.499	0.567	$n = 200$	SQ		0.147	0.052	0.024	0.306	0.736
	$\tilde{T}_n^\rho$		0.774	0.490	0.061	0.731	0.862		$\tilde{T}_n^\rho$		0.522	0.148	0.070	0.616	0.966
$n = 400$	SQ		0.665	0.512	0.029	0.633	0.667	$n = 400$	SQ		0.235	0.095	0.028	0.422	0.809
	$\tilde{T}_n^\rho$		0.937	0.841	0.055	0.938	0.928		$\tilde{T}_n^\rho$		0.824	0.395	0.052	0.900	0.996

As the tail distribution of the innovation process becomes heavier, the powers of the SQ-test (see the cases of $b_2 \neq 0.3(= b_1)$) become slightly worth. On the other hand, the GEL test keeps higher power than SQ-test even when the model has heavy-tails. Thus, we observed the advantage of the self-weighted GEL test.

5.5 Auxiliary Results

This section gives the proofs of Lemmas 5.1, 5.2 and supporting lemmas for the main results. Throughout this section, we assume all conditions of Theorem 5.5.

Proof of Lemma 5.1. Let us define

$$h_t(\theta, \tau) = \sup_{\|\theta' - \theta\| < \tau} \left\| g_t^*(\theta') - g_t^*(\theta) \right\|$$

and show that $h_t(\theta, \tau) \to 0$ almost surely as $\tau \to 0$. However, it suffices to show that $g_t^*(\theta)$ is continuous at each θ with probability one, and from the definition of $\varepsilon(t; \theta)$, $\text{sign}\{\varepsilon(t; \theta)\}$ is continuous at each θ with probability one. Hence, we get

$h_t(\theta, \tau) \to 0$ almost surely as $\tau \to 0$. Thus by the dominated convergence, for any ε and each θ, there exists $\tau(\theta)$ such that $E[h_t(\theta, \tau)] \le \varepsilon/4$ for all $\tau \le \tau(\theta)$. Next, define $B(\theta, \tau) = \{v \in \Theta : \|v - \theta\| < \tau\}$. By the compactness, there exist $\theta_1, ..., \theta_K$ such that $\{B(\theta_1, \tau(\theta_1)), ..., B(\theta_K, \tau(\theta_K))\}$ is a finite open covering of Θ. Let $\tau_k = \tau(\theta_k)$ and $\mu_k = E[h_t(\theta_k, \tau_k)]$. By the definition of τ_k, it follows that $\mu_k \le \varepsilon/4$ for all $k = 1, ..., K$. Now, for any θ, without loss of generality, let B_k contain θ. Then

$$
\left\| \hat{g}_n^*(\theta) - \bar{g}_n^*(\theta) \right\| \le \left(\frac{1}{n-u} \sum_{t=u+1}^{n} \left\| g_t^*(\theta) - g_t^*(\theta_k) \right\| - \mu_k \right) + \mu_k
$$

$$
+ \left\| \frac{1}{n-u} \sum_{t=u+1}^{n} g_t^*(\theta_k) - \bar{g}_n^*(\theta_k) \right\| + \left\| \bar{g}_n^*(\theta_k) - \bar{g}_n^*(\theta) \right\|
$$

$$
\le \left(\frac{1}{n-u} \sum_{t=u+1}^{n} h_t(\theta_k, \tau_k) - \mu_k \right) + \mu_k
$$

$$
+ \left\| \frac{1}{n-u} \sum_{t=u+1}^{n} g_t^*(\theta_k) - \bar{g}_n^*(\theta_k) \right\| + \left\| \bar{g}_n^*(\theta_k) - \bar{g}_n^*(\theta) \right\|.
$$

By the ergodicity, $(n-u)^{-1} \sum_{t=u+1}^{n} h_t(\theta_k, \tau_k) \xrightarrow{\text{a.s.}} \mu_k$ as $n \to \infty$. Therefore, for any $\varepsilon > 0$, there exists $n_{1k}(\varepsilon) \in \mathbb{N}$ such that $\|(n-u)^{-1} \sum_{t=u+1}^{n} h_i(\theta_k, \tau_k) - \mu_k\| \le \varepsilon/4$ a.s. for all $n \ge n_{1k}(\varepsilon)$. Similarly, there exists $n_{2k}(\varepsilon) \in \mathbb{N}$ such that $\|(n-u)^{-1} \sum_{t=u+1}^{n} g_t^*(\theta_k) - \bar{g}_n^*(\theta_k)\| \le \varepsilon/4$ a.s. for all $n \ge n_{2k}(\varepsilon)$. Furthermore,

$$
\left\| \bar{g}_n^*(\theta_k) - \bar{g}_n^*(\theta) \right\| = \left\| E[g_t^*(\theta_k) - g_t^*(\theta)] \right\|
$$

$$
\le E[\left\| g_t^*(\theta_k) - g_t^*(\theta) \right\|]
$$

$$
\le E\left[\sup_{\|\theta - \theta_k\| < \tau_k} \left\| g_t^*(\theta_k) - g_t^*(\theta) \right\| \right]
$$

$$
= E[h_t(\theta_k, \tau_k)]
$$

$$
= \mu_k
$$

$$
\le \varepsilon/4.
$$

Finally, let $n_k(\varepsilon) = \max\{n_{1k}(\varepsilon), n_{2k}(\varepsilon)\}$. Then for any $\varepsilon > 0$, it holds that

$$
\left\| \hat{g}_n^*(\theta) - \bar{g}_n^*(\theta) \right\| \le \varepsilon \quad \text{for all } n \ge n_k(\varepsilon) \text{ a.s.}
$$

Thus, we obtain

$$
\sup_{\theta \in \Theta} \left\| \hat{g}_n^*(\theta) - \bar{g}_n^*(\theta) \right\| \le \varepsilon \quad \text{for all } n \ge n(\varepsilon) \text{ a.s.,}
$$

where $n(\varepsilon) = \max\{n_k(\varepsilon) : k = 1, ..., K\}$. Hence, we get the desired result. $\square$

To prove Lemma 5.2, we impose Lemmas 5.3–5.8 below. The following lemma is equivalent to Assumption 2.2 (d) of Parente and Smith (2011).

Lemma 5.3 *For any $\tau_n \to 0$ as $n \to \infty$,*

$$\sup_{\|\theta - \theta_0\| \le \tau_n} \frac{n^{1/2}\|\hat{g}_n^*(\theta) - \hat{g}_n^*(\theta_0) - \bar{g}_n^*(\theta)\|}{1 + n^{1/2}\|\theta - \theta_0\|} = o_p(1).$$

Proof The proof is almost similar as Parente and Smith (2011, Proof of Theorem E.2, page 113), so we omit the proof here. □

On the other hand, the following lemmas are essentially due to Newey and Smith (2004).

Lemma 5.4 *Let $\Lambda_n = \{\lambda \in \mathbb{R}^m : \|\lambda\| \le c_0 n^{-1/2}\}$ for some $c_0 \in (0, \infty)$. Then,*

$$\sup_{\theta \in \Theta, \lambda \in \Lambda_n} \max_{u+1 \le t \le n} \left|\lambda^{\mathrm{T}} g_t^*(\theta)\right| = o_p(1)$$

and w.p.a.1, $\Lambda_n \subset \hat{\Lambda}_n(\theta)$ for all $\theta \in \Theta$.

Proof We first show that the quantity $\|w_{t-1}A_{t-1}(\theta)\|$ is bounded by a constant which is independent of t uniformly in $\theta \in \Theta$. Denote the lth element of $A_{t-1}(\theta)$ by $A_{l,t-1}(\theta)$ $(l = 1, \ldots, m)$ hereafter. From Assumption 5.1, we have, for $l = 1, \ldots, p$,

$$A_{l,t-1}(\theta) = -\psi(B; a)^{-1}\tilde{X}(t-l) = \sum_{k=0}^{\infty} \kappa_k^{(1)}(a)\tilde{X}(t-l-k) = \sum_{k=0}^{t-l-1} \kappa_k^{(1)}(a)X(t-l-k),$$

where $\{\kappa_k^{(1)}(a) : k \in \mathbb{Z}\}$ satisfy $|\kappa_{k-l}^{(1)}(a)| \le c_l r_l^k$ for some $c_l > 0$ and $r_l \in (0, 1)$ uniformly in a. Therefore, we have

$$\left|w_{t-1}^{1/2}A_{l,t-1}(\theta)\right| \le \frac{\sum_{k=0}^{t-l-1}|\kappa_k^{(1)}(a)||X(t-l-k)|}{1 + \sum_{k=1}^{t-1}k^{-\gamma}|X(t-k)|}$$

$$= \frac{\sum_{k=l}^{t-1}|\kappa_{k-l}^{(1)}(a)||X(t-k)|}{1 + \sum_{k=1}^{t-1}k^{-\gamma}|X(t-k)|}$$

$$\le \frac{\sum_{k=l}^{t-1}|\kappa_{k-l}^{(1)}(a)||X(t-k)|}{1 + \sum_{k=l}^{t-1}k^{-\gamma}|X(t-k)|}$$

$$\le c_l \frac{\sum_{k=l}^{t-1}r_l^k|X(t-k)|}{1 + \sum_{k=l}^{t-1}k^{-\gamma}|X(t-k)|} \quad \text{(by } |\kappa_{k-l}^{(1)}(a)| \le c_l r_l^k)$$

$$\le c_l \sum_{k=1}^{\infty} k^{\gamma} r_l^k, \tag{5.22}$$

where the right-hand side of (5.22) is independent of t, and $w_{t-1}^{1/2} \leq 1$. Therefore, it is shown that $|w_{t-1}A_{l,t-1}(\theta)|$ is bounded by a constant which is independent of t uniformly in $\theta \in \Theta$. By the same argument and the definition (5.3), it is also shown that $|w_{t-1}A_{p+j,t-1}(\theta)|$ is bounded by a constant for all $j = 1, \ldots, q$ uniformly in $\theta \in \Theta$ and t. Furthermore, by Assumption 5.2, $\|w_{t-1}\varphi_{t-1}\|$ is bounded by a constant which is independent of t. Hence, $\|g_t^*(\theta)\|$ is bounded by a constant which is independent of t and θ. Therefore, we get

$$\sup_{\theta \in \Theta, \lambda \in \Lambda_n} \max_{u+1 \leq t \leq n} \left| \lambda^{\mathrm{T}} g_t^*(\theta) \right| \leq Cn^{-1/2} = o_p(1),$$

so w.p.a.1, $\lambda^{\mathrm{T}} g_t^*(\theta) \in \mathcal{V}_\rho$ for all $\theta \in \Theta$ and $\|\lambda\| \leq c_0 n^{-1/2}$.   □

Lemma 5.5 *Suppose that there exists $\bar{\theta} \in \Theta$ such that $\bar{\theta} \to \theta_0$ in probability. Then,*

$$\left\| \frac{1}{n-u} \sum_{t=u+1}^{n} g_t^*(\bar{\theta})g_t^*(\bar{\theta})^{\mathrm{T}} - \Omega \right\| = o_p(1).$$

Proof By the definition, we can write $(n-u)^{-1} \sum_{t=u+1}^{n} g_t^*(\bar{\theta})g_t^*(\bar{\theta})^{\mathrm{T}}$ as

$$\frac{1}{n-u} \sum_{t=u+1}^{n} g_t^*(\bar{\theta})g_t^*(\bar{\theta})^{\mathrm{T}} = \frac{1}{n-u} \sum_{t=u+1}^{n} w_{t-1}^2 \begin{pmatrix} A_{t-1}(\bar{\theta})A_{t-1}(\bar{\theta})^{\mathrm{T}} & A_{t-1}(\bar{\theta})\varphi_{t-1}^{\mathrm{T}} \\ \varphi_{t-1}A_{t-1}(\bar{\theta})^{\mathrm{T}} & \varphi_{t-1}\varphi_{t-1}^{\mathrm{T}} \end{pmatrix}.$$
$$(5.23)$$

We shall show the consistency of each submatrix in (5.23) in succession.

First, we focus on the (l, j)th element of the first $m \times m$-submatrix of (5.23). For simplicity, we adopt the notation $\bar{A}_{l,t-1} = A_{l,t-1}(\bar{\theta})$ and $A_{l,t-1}^0 = A_{l,t-1}(\theta_0)$. Then, we have the following decomposition:

$$\frac{1}{n-u} \sum_{t=u+1}^{n} w_{t-1}^2 \bar{A}_{l,t-1}\bar{A}_{j,t-1} = \left(\bar{\Omega}_{n,A} - \Omega_{n,A}\right) + \left(\Omega_{n,A} - \Omega_{n,S}\right) + \Omega_{n,S}, \quad (5.24)$$

where

$$\bar{\Omega}_{n,A} = \frac{1}{n-u} \sum_{t=u+1}^{n} \tilde{w}_{t-1}^2 \bar{A}_{i,t-1}\bar{A}_{j,t-1},$$

$$\Omega_{n,A} = \frac{1}{n-u} \sum_{t=u+1}^{n} \delta_{t-1}^2 A_{i,t-1}^0 A_{j,t-1}^0,$$

$$\Omega_{n,S} = \frac{1}{n-u} \sum_{t=u+1}^{n} \delta_{t-1}^2 S_{i,t-1}S_{j,t-1}$$

and $S_{i,t-1}$ is the ith element of $S_{t-1} = (U_{t-1}, \ldots, U_{t-p}, V_{t-1}, \ldots, V_{t-q})^{\mathrm{T}}$.

For the first part of (5.24), the expansion $\bar{A}_{l,t-1} = A^0_{l,t-1} + (\partial_\theta \bar{A}^0_{l,t-1})^{\mathrm{T}}(\bar{\theta} - \theta_0)$ holds, where $\partial_\theta \bar{A}^0_{l,t-1} = (\partial A_{l,t-1}(\theta)/\partial \theta)|_{\theta=\bar{\theta}_0}$, and $\bar{\theta}_0$ is on the line joining $\bar{\theta}$ and θ_0. So the first term of (5.24) is decomposed as

$$\bar{\Omega}_{n,A} - \Omega_{n,A} = \frac{1}{n-u} \sum_{t=u+1}^{n} \delta_{t-1}^2 \left(\bar{A}_{l,t-1} \bar{A}_{j,t-1} - A^0_{l,t-1} A^0_{j,t-1} \right)$$

$$+ \frac{1}{n-u} \sum_{t=u+1}^{n} \left(\tilde{w}_{t-1}^2 - \delta_{t-1}^2 \right) \bar{A}_{l,t-1} \bar{A}_{j,t-1}$$

$$= \frac{1}{n-u} \sum_{t=u+1}^{n} \delta_{t-1}^2 \left(A^0_{l,t-1} (\partial_\theta \bar{A}^0_{j,t-1})^{\mathrm{T}} + A^0_{j,t-1} (\partial_\theta \bar{A}^0_{l,t-1})^{\mathrm{T}} \right)(\bar{\theta} - \theta_0)$$

$$\tag{5.25}$$

$$+ (\bar{\theta} - \theta_0)^{\mathrm{T}} \frac{1}{n-u} \sum_{t=u+1}^{n} \delta_{t-1}^2 (\partial_\theta \bar{A}^0_{l,t-1})(\partial_\theta \bar{A}^0_{j,t-1})^{\mathrm{T}}(\bar{\theta} - \theta_0) \tag{5.26}$$

$$+ \frac{1}{n-u} \sum_{t=u+1}^{n} \left(\tilde{w}_{t-1}^2 - \delta_{t-1}^2 \right) \bar{A}_{l,t-1} \bar{A}_{j,t-1}. \tag{5.27}$$

By the similar argument as in the proof of Lemma 5.4, the summands in (5.25) and (5.26) are bounded by some constants with probability one. From this fact and $\bar{\theta} - \theta_0 \to 0$ in probability , the terms (5.25) and (5.26) converge to zero in probability as $n \to \infty$. On the other hand, we have

$$\left| \frac{1}{n-u} \sum_{t=u+1}^{n} \left(\tilde{w}_{t-1}^2 - \delta_{t-1}^2 \right) \bar{A}_{l,t-1} \bar{A}_{j,t-1} \right|$$

$$\leq \frac{1}{n-u} \sum_{t=u+1}^{n} |\tilde{w}_{t-1} - \delta_{t-1}||\tilde{w}_{t-1} + \delta_{t-1}| \left| \bar{A}_{l,t-1} \right| \left| \bar{A}_{j,t-1} \right|$$

$$\leq \frac{2}{n-u} \sum_{t=u+1}^{n} |\tilde{w}_{t-1} - \delta_{t-1}| \left| \tilde{w}_{t-1}^{1/2} \bar{A}_{l,t-1} \right| \left| \tilde{w}_{t-1}^{1/2} \bar{A}_{j,t-1} \right| \quad (\text{by } \delta_{t-1} \leq w_{t-1} \leq 1)$$

$$\leq \frac{C}{n-u} \sum_{t=u+1}^{n} |\tilde{w}_{t-1} - \delta_{t-1}| \to 0 \; (\text{in probability}). \tag{5.28}$$

Therefore, the term (5.27) converges to zero in probability as $n \to \infty$.

For the second part of (5.24), we have $|A^0_{l,t-1} - S_{l,t-1}| \leq \xi_t$ from Lemma 1 of Pan et al. (2007), where $\xi_t = c \sum_{k=t}^{\infty} r^k |y_{t-k}|$ for some $c' \in (0, \infty)$ and $r \in (0, 1)$. Obviously, $\xi_t = o_p(1)$ as $t \to \infty$ and hence $\Omega_{n,A} - \Omega_{n,S} = o_p(1)$.

For the third part of (5.24), it is easy to see that $\Omega_{n,S}$ converges to the first $m \times m$-submatrix of Ω by the ergodicity of S_{t-1}.

Second, we consider the last $d \times d$-submatrix of (5.23). For $l, j \in \{1, \ldots, d\}$, consider the decomposition

$$\frac{1}{n-u} \sum_{t=1}^{n} \tilde{w}_{t-1}^2 \varphi_{l,t-1} \varphi_{j,t-1} = \frac{1}{n-u} \sum_{t=1}^{n} \delta_{t-1}^2 \varphi_{l,t-1} \varphi_{j,t-1} \tag{5.29}$$

$$+ \frac{1}{n-u} \sum_{t=1}^{n} (\tilde{w}_{t-1}^2 - \delta_{t-1}^2) \varphi_{l,t-1} \varphi_{j,t-1}. \tag{5.30}$$

Note that (5.29) converges to $E[\delta_{t-1}^2 \varphi_{l,t-1} \varphi_{j,t-1}]$ a.e. from

$$E[|\delta_{t-1}^2 \varphi_{l,t-1} \varphi_{j,t-1}|] \leq E[|\tilde{w}_{t-1}^2 \varphi_{l,t-1} \varphi_{j,t-1}|] < \infty$$

by Assumption 5.2, stationarity and ergodicity of $\delta_{t-1}^2 \varphi_{l,t-1} \varphi_{j,t-1}$. On the other hand, it is shown that (5.30) converges to zero in probability as $n \to \infty$ by the same argument as (5.28) and Assumption 5.2.

Third, we show the consistency of the off-diagonal part of (5.23). For $l \in \{1, \ldots, m\}$ and $j \in \{1, \ldots, d\}$, we have

$$\frac{1}{n-u} \sum_{t=u+1}^{n} \tilde{w}_{t-1}^2 \bar{A}_{l,t-1} \varphi_{j,t-1} = \frac{1}{n-u} \sum_{t=u+1}^{n} \delta_{t-1}^2 A_{l,t-1}^0 \varphi_{j,t-1} \tag{5.31}$$

$$+ \frac{1}{n-u} \sum_{t=u+1}^{n} (\tilde{w}_{t-1}^2 - \delta_{t-1}^2) A_{l,t-1}^0 \varphi_{j,t-1} \tag{5.32}$$

$$+ \frac{1}{n-u} \sum_{t=u+1}^{n} \tilde{w}_{t-1}^2 (\partial_\theta \bar{A}_{l,t-1}^0)^{\mathrm{T}} (\bar{\theta} - \theta_0) \varphi_{j,t-1}. \tag{5.33}$$

Again from Lemma 1 of Pan et al. (2007), $(n-u)^{-1} \sum_{t=u+1}^{n} \delta_{t-1}^2 \{A_{l,t-1}^0 - S_{l,t-1}\} \varphi_{j,t-1} = o_p(1)$ and hence (5.31) converges to $E[\delta_{t-1}^2 S_{l,t-1} \varphi_{j,t-1}]$ in probability. On the other hand, (5.32) and (5.33) converge to zero in probability by the Cauchy–Schwarz inequality and the same arguments above. Thus, we get the desired result. $\qquad \square$

Lemma 5.6 *Suppose that there exists $\bar{\theta} \in \Theta$ such that $\bar{\theta} \to \theta_0$ in probability and $\hat{g}^*(\bar{\theta}) = O_p(n^{-1/2})$. Then,*

$$\bar{\lambda} = \arg \max_{\lambda \in \bar{\Lambda}_n(\bar{\theta})} P_n^*(\bar{\theta}, \lambda)$$

exists w.p.a.1, $\bar{\lambda} = O_p(n^{-1/2})$ and $P_n^*(\bar{\theta}, \bar{\lambda}) = O_p(1)$.

Proof Since Λ_n is a closed set, $\bar{\lambda} = \arg \max_{\lambda \in \Lambda_n} P_n^*(\bar{\theta}, \lambda)$ exists with probability one. From Lemma 5.4, $P_n^*(\bar{\theta}, \lambda)$ is continuously twice differentiable w.p.a.1 with respect to λ. So by a Taylor expansion around $\lambda = 0_m$, there exists $\check{\lambda}$ on the line joining $\check{\lambda}$ and 0_m such that

$$
\begin{aligned}
0 &= P_n^*(\bar{\theta}, 0_m) \\
&\leq P_n^*(\bar{\theta}, \check{\lambda}) \\
&= -n\check{\lambda}^{\mathrm{T}} \hat{g}^*(\bar{\theta}) + \frac{n}{2} \check{\lambda}^{\mathrm{T}} \left[\frac{1}{n-u} \sum_{t=u+1}^{n} \bar{\rho}_t' g_t^*(\bar{\theta}) g_t^*(\bar{\theta})^{\mathrm{T}} \right] \check{\lambda},
\end{aligned}
\tag{5.34}
$$

where $\bar{\rho}_t' = \rho'\{\check{\lambda}^{\mathrm{T}} g_t^*(\bar{\theta})\}$. Furthermore, by Lemmas 5.4 and 5.5,

$$
\bar{\Omega}_n^\rho = -\frac{1}{n-u} \sum_{t=u+1}^{n} \bar{\rho}_t' g_t^*(\bar{\theta}) g_t^*(\bar{\theta})^{\mathrm{T}} \to \Omega
$$

in probability, and therefore the minimum eigenvalue of $\bar{\Omega}_n^\rho$ is bounded away from 0 w.p.a.1. from (ii) of Assumption 5.4. Then, it holds that

$$
0 \leq -\check{\lambda}^{\mathrm{T}} \hat{g}^*(\bar{\theta}) - \frac{1}{2} \check{\lambda}^{\mathrm{T}} \bar{\Omega}_n^\rho \check{\lambda} \leq \|\check{\lambda}\| \|\hat{g}^*(\bar{\theta})\| - C\|\check{\lambda}\|^2
\tag{5.35}
$$

w.p.a.1. Dividing both side of (5.35) by $\|\check{\lambda}\|$, we get $\|\check{\lambda}\| = O_p(n^{-1/2})$, and hence $\check{\lambda} \in \Lambda_n$ w.p.a.1. Again by Lemma 5.4, concavity of $P_n^*(\bar{\theta}, \lambda)$ and convexity of $\hat{\Lambda}_n(\bar{\theta})$, it is shown that $\bar{\lambda} = \check{\lambda}$ exists w.p.a.1 and $\bar{\lambda} = O_p(n^{-1/2})$. These results and (5.34) for $\check{\lambda} = \bar{\lambda}$ also imply that $P_n^*(\bar{\theta}, \bar{\lambda}) = O_p(1)$. $\square$

Lemma 5.7 $\hat{g}^*(\hat{\theta}^*) = O_p(n^{-1/2})$ as $n \to \infty$.

Proof We define $\hat{\hat{g}} = \hat{g}^*(\hat{\theta}^*)$ and $\tilde{\lambda} = -n^{-1/2} \hat{\hat{g}} / \|\hat{\hat{g}}\|$.
    First, by a quite similar argument as in the proof of Lemma 5.6, and by noting that $\rho'\{\tilde{\lambda}^{\mathrm{T}} g_t^*(\theta)\} \geq -C$ uniformly in t and θ w.p.a.1. from Lemma 5.4, we have

$$
\begin{aligned}
P_n^*(\hat{\theta}^*, \tilde{\lambda}) &\geq n \left(n^{-1/2} \|\hat{\hat{g}}\| - \frac{C}{2} \tilde{\lambda}^{\mathrm{T}} \left[\frac{1}{n-u} \sum_{t=u+1}^{n} g_t^*(\hat{\theta}^*) g_t^*(\hat{\theta}^*)^{\mathrm{T}} \right] \tilde{\lambda} \right) \\
&\geq n \left(n^{-1/2} \|\hat{\hat{g}}\| - C\|\tilde{\lambda}\|^2 \right) \\
&= \left(n^{1/2} \|\hat{\hat{g}}\| - C \right)
\end{aligned}
\tag{5.36}
$$

w.p.a.1. Second, by the definition of $\hat{\lambda}^*$,

$$
P_n^*(\hat{\theta}^*, \hat{\lambda}^*) = \max_{\lambda \in \hat{\Lambda}_n(\hat{\theta})} P_n^*(\hat{\theta}^*, \lambda) \geq P_n^*(\hat{\theta}^*, \tilde{\lambda}).
\tag{5.37}
$$

Third, note that the central limit theorem yields $\hat{g}^*(\theta_0) = O_p(n^{-1/2})$. Then by applying Lemma 5.6 for $\bar{\theta} = \theta_0$, we get $\sup_{\lambda \in \hat{\Lambda}_n(\theta_0)} P_n^*(\theta_0, \lambda) = O_p(1)$. Thus we obtain

$$P_n^*(\hat{\theta}^*, \hat{\lambda}^*) = \min_{\theta \in \Theta} \sup_{\lambda \in \hat{\Lambda}_n(\theta)} P_n^*(\theta, \lambda) \leq \sup_{\lambda \in \hat{\Lambda}_n(\theta_0)} P_n^*(\theta_0, \lambda) = O_p(1). \tag{5.38}$$

Finally, from (5.36)–(5.38), $n^{1/2}\|\hat{\hat{g}}\| = O_p(1)$, which implies the assertion of this lemma.   □

Lemma 5.8 $\hat{\theta}^* - \theta_0 = O_p(n^{-1/2})$.

Proof It follows from the triangular inequality, Lemmas 5.1 and 5.7 that

$$\|\bar{g}_n^*(\hat{\theta}^*)\| \leq \|\bar{g}_n^*(\hat{\theta}^*) - \hat{g}_n^*(\hat{\theta}^*)\| + \|\hat{g}_n^*(\hat{\theta}^*)\|$$
$$\leq \sup_{\theta \in \Theta} \|\bar{g}_n^*(\theta) - \hat{g}_n^*(\theta)\| + \|\hat{g}_n^*(\hat{\theta}^*)\| = o_p(1).$$

Since $\bar{g}_n^*(\theta) - E[g_t^{*0}(\theta)] = o_p(1)$ uniformly in θ and $E[g_t^{*0}(\theta)]$ has a unique zero at θ_0 by Assumption 5.4, $\|\bar{g}_n^*(\theta)\|$ must be bounded away from zero outside any neighborhood of θ_0. Therefore, $\hat{\theta}^*$ must be inside any neighborhood of θ_0 w.p.a.1. Then, $\hat{\theta}^* \to \theta$ in probability .

Next, we show that $\hat{\theta}^* - \theta_0 = O_p(n^{-1/2})$. By Lemma 5.7, $\hat{g}_n^*(\hat{\theta}^*) = O_p(n^{-1/2})$ and by the central limit theorem, $\hat{g}_n^*(\theta_0)$ is also $O_p(n^{-1/2})$. Further, from Lemma 5.3,

$$\|\hat{g}_n^*(\hat{\theta}^*) - \hat{g}_n^*(\theta_0) - \bar{g}_n^*(\hat{\theta}^*)\| \leq (1 + n^{1/2}\|\hat{\theta}^* - \theta_0\|)o_p(n^{-1/2}). \tag{5.39}$$

Therefore,

$$\|\bar{g}_n^*(\hat{\theta}^*)\| \leq \|\hat{g}_n^*(\hat{\theta}^*) - \hat{g}_n^*(\theta_0) - \bar{g}_n^*(\hat{\theta}^*)\| + \|\hat{g}_n^*(\hat{\theta}^*)\| + \|\hat{g}_n^*(\theta_0)\|$$
$$= (1 + n^{1/2}\|\hat{\theta}^* - \theta_0\|)o_p(n^{-1/2}) + O_p(n^{-1/2}).$$

In addition, by the similar argument as Newey and McFadden (1994, p. 2191) and differentiability of $\|\bar{g}_n^*(\theta)\|$, $\|\bar{g}_n^*(\hat{\theta}^*)\| \geq C\|\hat{\theta}^* - \theta_0\|$ w.p.a.1. Therefore, we get

$$\|\hat{\theta}^* - \theta_0\| = (1 + n^{1/2}\|\hat{\theta}^* - \theta_0\|)o_p(n^{-1/2}) + O_p(n^{-1/2})$$

and hence $\|\hat{\theta}^* - \theta_0\| = O_p(n^{-1/2})/\{1 + o_p(1)\} = O_p(n^{-1/2})$.   □

Now, we get the proof of Lemma 5.2 as follows.

Proof of Lemma 5.2. It is suffice to show the following three relationships:

(i) $P_n^*(\hat{\theta}^*, \hat{\lambda}^*) - L_n^*(\hat{\theta}^*, \hat{\lambda}^*) = o_p(1)$,
(ii) $L_n^*(\hat{\theta}^*, \hat{\lambda}^*) - L_n^*(\hat{\theta}^L, \hat{\lambda}^*) = o_p(1)$,
(iii) $L_n^*(\hat{\theta}^L, \hat{\lambda}^*) - L_n^*(\hat{\theta}^L, \hat{\lambda}^L) = o_p(1)$.

For (i), Taylor expansion yields

$$P_n^*(\hat{\theta}^*, \hat{\lambda}^*) = -n\hat{\lambda}^{*\mathrm{T}}\hat{g}^*(\hat{\theta}^*) + \frac{n}{2}\hat{\lambda}^{*\mathrm{T}}\left[\frac{1}{n-u}\sum_{t=u+1}^{n}\rho'\{\bar{\lambda}^{\mathrm{T}}g_t^*(\hat{\theta}^*)\}g_t^*(\hat{\theta}^*)g_t^*(\hat{\theta}^*)^{\mathrm{T}}\right]\hat{\lambda}^*,$$

where $\bar{\lambda}$ is on the line joining 0_m and $\hat{\lambda}^*$. Then

$$\left|P_n^*(\hat{\theta}^*, \hat{\lambda}^*) - \hat{L}_n(\hat{\theta}^*, \hat{\lambda}^*)\right| \leq \left|-n\left(\hat{g}^*(\hat{\theta}^*) - \hat{g}^*(\theta_0) - G(\hat{\theta}^* - \theta_0)\right)^{\mathrm{T}}\hat{\lambda}^*\right| \tag{5.40}$$

$$+\left|\frac{n}{2}\hat{\lambda}^{*\mathrm{T}}\left[\frac{1}{n-u}\sum_{t=u+1}^{n}\rho'\{\bar{\lambda}^{\mathrm{T}}g_t^*(\hat{\theta}^*)\}g_t^*(\hat{\theta}^*)g_t^*(\hat{\theta}^*)^{\mathrm{T}} + \Omega\right]\hat{\lambda}^*\right|. \tag{5.41}$$

Since $\hat{\theta}^* \to \theta_0$ in probability by Lemma 5.8, we can apply Lemma 5.6 for $\bar{\theta} = \hat{\theta}^*$ and hence $\hat{\lambda}^* = O_p(n^{-1/2})$. Then, by recalling (5.39), the quantity (5.40) becomes

$$\left|-n\left(\hat{g}^*(\hat{\theta}^*) - \hat{g}^*(\theta_0) - G(\hat{\theta}^* - \theta_0)\right)^{\mathrm{T}}\hat{\lambda}^*\right|$$

$$\leq n\left\{\left\|\hat{g}^*(\hat{\theta}^*) - \hat{g}^*(\theta_0) - g^*(\hat{\theta}^*)\right\| + \left\|g^*(\hat{\theta}^*) - G(\hat{\theta}^* - \theta_0)\right\|\right\}\left\|\hat{\lambda}^*\right\|$$

$$= \left\{\left(1 + n^{1/2}\left\|\hat{\theta}^* - \theta_0\right\|\right)o_p(n^{-1/2}) + O_p\left(\left\|\hat{\theta}^* - \theta_0\right\|^2\right)\right\}O_p(n^{1/2})$$

$$= o_p(1).$$

Moreover, (5.41) is $o_p(1)$. Hence, we get $\left|P_n^*(\hat{\theta}^*, \hat{\lambda}^*) - L_n^*(\hat{\theta}^*, \hat{\lambda}^*)\right| = o_p(1)$.

To get (ii), we first show $|P_n^*(\hat{\theta}^L, \hat{\lambda}^*) - L_n^*(\hat{\theta}^L, \hat{\lambda}^*)| = o_p(1)$. Note that $L_n^*(\theta, \lambda)$ is smooth in θ and λ. Then, the first-order conditions for an interior global maximum,

$$0_d = \frac{\partial L_n^*(\theta, \lambda)}{\partial \theta} = -nG^{\mathrm{T}}\lambda, \tag{5.42}$$

$$0_m = \frac{\partial L_n^*(\theta, \lambda)}{\partial \lambda} = -n\left\{G(\theta - \theta_0) + \hat{g}^*(\theta_0) + \Omega\lambda\right\} \tag{5.43}$$

are satisfied at $(\hat{\theta}^{L\mathrm{T}}, \hat{\lambda}^{L\mathrm{T}})^{\mathrm{T}}$. The conditions above are stacked as

$$\begin{pmatrix} O_{d\times d} & G^{\mathrm{T}} \\ G & \Omega \end{pmatrix}\begin{pmatrix} \hat{\theta}^L - \theta_0 \\ \hat{\lambda}^L \end{pmatrix} + \begin{pmatrix} 0_d \\ \hat{g}^*(\theta_0) \end{pmatrix} = 0_{d+m}. \tag{5.44}$$

By denoting $\Sigma = (G^{\mathrm{T}}\Omega^{-1}G)^{-1}$, $H = \Omega^{-1}G\Sigma$ and $P = \Omega^{-1} - H\Sigma^{-1}H^{\mathrm{T}}$, (5.44) is equivalent to

$$\begin{pmatrix} \hat{\theta}^L - \theta_0 \\ \hat{\lambda}^L \end{pmatrix} = - \begin{pmatrix} \Sigma & -H^T \\ -H & -P \end{pmatrix} \begin{pmatrix} 0_d \\ -\hat{g}^*(\theta_0) \end{pmatrix} = \begin{pmatrix} -H^T \hat{g}^*(\theta_0) \\ -P \hat{g}^*(\theta_0) \end{pmatrix}, \tag{5.45}$$

so both $\hat{\theta}^L - \theta_0$ and $\hat{\lambda}^L$ are $O_p(n^{-1/2})$. Therefore, by the same arguments as (i) in this proof, $|P_n^*(\hat{\theta}^L, \hat{\lambda}^*) - L_n^*(\hat{\theta}^L, \hat{\lambda}^*)| = o_p(1)$. This relationship and the fact that $(\hat{\theta}^T, \hat{\lambda}^{*T})^T$ and $(\hat{\theta}^{LT}, \hat{\lambda}^{LT})^T$ are, respectively, the saddle points of $P_n^*(\theta, \lambda)$ and $L_n^*(\theta, \lambda)$ imply that

$$L_n^*(\hat{\theta}^*, \hat{\lambda}^*) = P_n^*(\hat{\theta}^*, \hat{\lambda}^*) + o_p(1) \le P_n^*(\hat{\theta}^L, \hat{\lambda}^*) + o_p(1) = L_n^*(\hat{\theta}^L, \hat{\lambda}^*) + o_p(1). \tag{5.46}$$

On the other hand,

$$\begin{aligned} L_n^*(\hat{\theta}^L, \hat{\lambda}^*) &\le L_n^*(\hat{\theta}^L, \hat{\lambda}^L) \\ &\le L_n^*(\hat{\theta}^*, \hat{\lambda}^L) = P_n^*(\hat{\theta}^*, \hat{\lambda}^L) + o_p(1) \\ &\le P_n^*(\hat{\theta}^*, \hat{\lambda}^*) + o_p(1) = L_n^*(\hat{\theta}^*, \hat{\lambda}^*) + o_p(1). \end{aligned} \tag{5.47}$$

Thus, (5.46) and (5.47) yield $L_n^*(\hat{\theta}^*, \hat{\lambda}^*) - L_n^*(\hat{\theta}^L, \hat{\lambda}^*) = o_p(1)$.

We can prove (iii) similarly, i.e.,

$$\begin{aligned} \hat{L}_n(\hat{\theta}^L, \hat{\lambda}^L) &\le L_n^*(\hat{\theta}^*, \hat{\lambda}^L) = P_n^*(\hat{\theta}^*, \hat{\lambda}^L) + o_p(1) \\ &\le P_n^*(\hat{\theta}^*, \hat{\lambda}^*) + o_p(1) \\ &\le P_n^*(\hat{\theta}^L, \hat{\lambda}^*) + o_p(1) = L_n^*(\hat{\theta}^L, \hat{\lambda}^*) + o_p(1) \end{aligned}$$

and $L_n^*(\hat{\theta}^L, \hat{\lambda}^*) \le L_n^*(\hat{\theta}^L, \hat{\lambda}^L)$. That is, $L_n^*(\hat{\theta}^L, \hat{\lambda}^*) = L_n^*(\hat{\theta}^L, \hat{\lambda})^L + o_p(1)$. □

Bibliography

Akashi F (2017) Self-weighted generalized empirical likelihood methods for hypothesis testing in infinite variance arma models. Statist Inference Stoch Process 20(3):291–313

Akashi F, Dette H, Liu Y (2018) Change-point detection in autoregressive models with no moment assumptions. J Time Ser Anal 39(5):763–786

Akashi F, Liu Y, Taniguchi M (2015) An empirical likelihood approach for symmetric α-stable processes. Bernoulli 21(4):2093–2119

Albrecht V (1984) On the convergence rate of probability of error in Bayesian discrimination between two Gaussian processes. In: Proceedings of the third Prague symposium on asymptotic statistics

Basu A, Harris IR, Hjort NL, Jones M (1998) Robust and efficient estimation by minimising a density power divergence. Biometrika 85(3):549–559

Bingham NH, Goldie CM, Teugels JL (1987) Regular variation. Cambridge University Press

Birr S, Volgushev S, Kley T, Dette H, Hallin M (2017) Quantile spectral analysis for locally stationary time series. J Royal Statist Soc Ser B (Statist Methodol) 79(5):1619–1643

Bloomfield P (1970) Spectral analysis with randomly missing observations. J Royal Statist Soc Ser B (Methodol) 32(3):369–380

Bloomfield P (1973) An exponential model for the spectrum of a scalar time series. Biometrika 60(2):217–226

Box GE, Jenkins GM (1976) Time series analysis: forecasting and control. Holden-Day

Brillinger DR (2001) Time series: data analysis and theory. Vol. 36 SIAM

Brockwell PJ, Davis RA (1991) Time series: theory and methods. Springer

Chuang CS, Chan NH (2002) Empirical likelihood for autoregressive models, with applications to unstable time series. Statist Sin 12(2):387–407

Ciuperca G, Salloum Z (2015) Empirical likelihood test in a posteriori change-point nonlinear model. Metrika 78(8):919–952

Cressie N, Read TR (1984) Multinomial goodness-of-fit tests. J Royal Statist Soc Ser B (Methodol) 46(3):440–464

Csiszár I (1975) I-divergence geometry of probability distributions and minimization problems. Ann Probab 3(1):146–158

Csiszár I (1991) Why least squares and maximum entropy? an axiomatic approach to inference for linear inverse problems. Anna Statist 19(4):2032–2066

Davis R, Resnick S (1986) Limit theory for the sample covariance and correlation functions of moving averages. Annals of Statist 14(2):533–558

Dette H, Hallin M, Kley T, Volgushev S et al (2015) Of copulas, quantiles, ranks and spectra: an L_1-approach to spectral analysis. Bernoulli 21(2):781–831

Duren PL (1970) Theory of H^p spaces. IMA

Dzhaparidze K (1986) Parameter estimation and hypothesis testing in spectral analysis of stationary time series. Springer

Feller W (1968) An introduction to probability theory and its applications. Wiley

Fujisawa H, Eguchi S (2008) Robust parameter estimation with a small bias against heavy contamination. J Multivar Anal 99(9):2053–2081

Geyer CJ (1996) On the asymptotics of convex stochastic optimization. Unpublished manuscript

Grenander U, Rosenblatt M (1957) Statistical analysis of stationary time series. Wiley

Hall P, Heyde C (1980) Martingale limit theory and its application. Academic Press

Hamilton JD (1994) Time series analysis. Princeton University Press

Hannan EJ (1970) Multiple time series. Wiley, New York

Hannan EJ (1973) The estimation of frequency. J Appl Probab 10(3):510–519

Hansen LP, Heaton J, Yaron A (1996) Finite-sample properties of some alternative gmm estimators. J Bus Econ Statist 14(3):262–280

Helson H (1964) Lectures on invariant subspaces. Academic Press, New York

Helson H, Lowdenslager D (1958) Prediction theory and fourier series in several variables. Acta Mathematica 99(1):165–202

Hewitt E, Stromberg K (1975) Real and abstract analysis—a modern treatment of the theory of functions of a real variable. Vol. 25. Splinger

Hoffman K (1962) Banach spaces of analytic functions. Englewood Cliffs, NJ, Prentice-Hall

Hosoya Y (1978) Robust linear extrapolations of second-order stationary processes. Ann Probab 6(4):574–584

Hosoya Y (1997) A limit theory for long-range dependence and statistical inference on related models. Ann Statist 25(1):105–137

Hosoya Y, Taniguchi M (1982) A central limit theorem for stationary processes and the parameter estimation of linear processes. Annals Statist 10(1):132–153

Huber PJ (1967) The behavior of maximum likelihood estimates under nonstandard conditions. In: Proceedings of the fifth Berkeley symposium on mathematical statistics and probability

Hunt RA (1968) On the convergence of fourier series. In: Orthogonal expansions and their continuous analogues, pp 235–255

Keenan DM (1987) Limiting behavior of functionals of higher-order sample cumulant spectra. Ann Statist 15(1):134–151

Kholevo A (1969) On estimates of regression coefficients. Theor Probab Appl 14(1):79–104

Kitamura Y (1997) Empirical likelihood methods with weakly dependent processes. Ann Statist 25(5):2084–2102

Kitamura Y, Stutzer M (1997) An information-theoretic alternative to generalized method of moments estimation. Econometrica 65(4):861–874

Kley T, Volgushev S, Dette H, Hallin M et al (2016) Quantile spectral processes: asymptotic analysis and inference. Bernoulli 22(3):1770–1807

Klüppelberg C, Mikosch T (1993) Spectral estimates and stable processes. Stoch Process Appl 47(2):323–344

Klüppelberg C, Mikosch T (1994) Some limit theory for the self-normalized periodogram of stable processes. Scand J Statist 21(4):485–491

Klüppelberg C, Mikosch T (1996) The integrated periodogram for stable processes. Ann Statist 24(5):1855–1879

Knight K (1998) Limiting distributions for L_1 regression estimators under general conditions. Ann Statist 26(2):755–770

Koenker R (2005) Quantile regression, vol 38. Cambridge University Press

Kolmogorov AN (1941a) Interpolation and extrapolation of stationary random sequences. Bull de l'academie des Sci de U.R.S.S Ser. Math 5:3–14

Kolmogorov AN (1941b) Stationary sequences in hilbert space. Bull Math Univ Moscow 2(6):1–40

Koosis P (1998) Introduction to H_p spaces. Cambridge University Press

Li J, Liang W, He S (2011) Empirical likelihood for lad estimators in infinite variance arma models. Statist Probab Let 81(2):212–219

Li TH (2008) Laplace periodogram for time series analysis. J Am Statist Assoc 103(482):757–768

Li TH (2012) Quantile periodograms. J Am Statist Assoc 107(498):765–776

Li TH, Kedem B, Yakowitz S (1994) Asymptotic normality of sample autocovariances with an application in frequency estimation. Stoch Process Appl 52(2):329–349

Ling S (2005) Self-weighted least absolute deviation estimation for infinite variance autoregressive models. J Royal Statist Soc Se B (Statist Methodol) 67(3):381–393

Liu Y (2017a) Robust parameter estimation for stationary processes by an exotic disparity from prediction problem. Statist Probab Lett 129:120–130

Liu Y (2017b) Statistical inference for quantiles in the frequency domain. Statist Inference Stoch Process 20(3):369–386

Liu Y, Xue Y, Taniguchi M (2018) Robust linear interpolation and extrapolation of stationary time series in L_p. Submited manuscript

Lütkepohl H (2005) New introduction to multiple time series analysis. Springer

Miamee A, Pourahmadi M (1988) Best approximations in $L^p(d\mu)$ and prediction problems of Szegö, Kolmogorov, Yaglom, and Nakazi. J Lond Math Soc 2(1):133–145

Mikosch T, Gadrich T, Klüppelberg C, Adler RJ (1995) Parameter estimation for arma models with infinite variance innovations. Ann Statist 23(1):305–326

Mikosch T, Resnick S, Samorodnitsky G (2000) The maximum of the periodogram for a heavy-tailed sequence. Ann Probab 28(2):885–908

Monti AC (1997) Empirical likelihood confidence regions in time series models. Biometrika 84(2):395–405

Newey WK (1991) Uniform convergence in probability and stochastic equicontinuity. Econom J Econom Soc 59(4):1161–1167

Newey WK, McFadden D (1994) Large sample estimation and hypothesis testing. Handb Econom 4:2111–2245

Newey WK, Smith RJ (2004) Higher order properties of GMM and generalized empirical likelihood estimators. Econometrica 72(1):219–255

Nolan JP (2012) Stable distributions—models for heavy tailed data. Birkhauser

Ogata H, Taniguchi M (2010) An empirical likelihood approach for non-Gaussian vector stationary processes and its application to minimum contrast estimation. Aust N Z J Statist 52(4):451–468

Owen AB (1988) Empirical likelihood ratio confidence intervals for a single functional. Biometrika 75(2):237–249

Pan J, Wang H, Yao Q (2007) Weighted least absolute deviations estimation for arma models with infinite variance. Econom Theor 23(5):852–879

Parente PM, Smith RJ (2011) GEL methods for nonsmooth moment indicators. Econom Theor 27(1):74–113

Petrov VV (1975) Sums of independent random variables. Springer

Pollard D (1991) Asymptotics for least absolute deviation regression estimators. Econom Theor 7(2):186–199

Qin J, Lawless J (1994) Empirical likelihood and general estimating equations. Ann Statist 22(1):300–325

Qu Z (2008) Testing for structural change in regression quantiles. J Econom 146(1):170–184

Quinn BG, Hannan EJ (2001) The estimation and tracking of frequency. vol 9 Cambridge University Press

Quinn BG, Thomson PJ (1991) Estimating the frequency of a periodic function. Biometrika 78(1):65–74

Rao CR, Mitra SK (1971) Generalized inverse of matrices and its applications. Wiley

Renyi A (1961) On measures of entropy and information. Fourth Berkeley Symp Math Statist Probab 1:547–561

Resnick SI, Stărică C (1996) Asymptotic behavior of hill's estimator for autoregressive data. Stoch Models 13(4):703–723

Resnick SI, Stărică C (1998) Tail index estimation for dependent data. Ann Appl Probab 8(4):1156–1183

Rice JA, Rosenblatt M (1988) On frequency estimation. Biometrika 75(3):477–484

Samoradnitsky G, Taqqu MS (1994) Stable non-gaussian random processes: stochastic models with infinite variance. vol 1. CRC Press

Suto Y, Liu Y, Taniguchi M (2016) Asymptotic theory of parameter estimation by a contrast function based on interpolation error. Statist Inference Stoch Process 19(1):93–110

Szegö G (1915) Ein grenzwertsatz über die toeplitzschen determinanten einer reellen positiven funktion. Math Ann 76:490–503

Taniguchi M (1981a) An estimation procedure of parameters of a certain spectral density model. J Royal Statist Soc Ser B 43(1):34–40

Taniguchi M (1981b) Robust regression and interpolation for time series. J Time Ser Anal 2(1):53–62

Taniguchi M (1982) On estimation of the integrals of the fourth order cumulant spectral density. Biometrika 69(1):117–122

Taniguchi M (1987) Minimum contrast estimation for spectral densities of stationary processes. J Royal Statist Soc Ser B 49(3):315–325

Taniguchi M, Kakizawa Y (2000) Asymptotic theory of statistical inference for time series. Springer

Taniguchi M, van Garderen KJ, Puri ML (2003) Higher order asymptotic theory for minimum contrast estimators of spectral parameters of stationary processes. Econom Theory 19(6):984–1007

Tauchen G (1985) Diagnostic testing and evaluation of maximum likelihood models. J Econom 30(1):415–443

Wald A (1939) Contributions to the theory of statistical estimation and testing hypotheses. Ann Math Statist 10(4):299–326

Wald A (1945) Statistical decision functions which minimize the maximum risk. Ann Math 46(2):265–280

Walker AM (1971) On the estimation of a harmonic component in a time series with stationary independent residuals. Biometrika 58(1):21–36

Whittle P (1952) Some results in time series analysis. Scand Actuar J 1952(1–2):48–60

Whittle P (1952b) Tests of fit in time series. Biometrika 39(3):309–318

Whittle P (1953) Estimation and information in stationary time series. Arkiv för matematik 2(5):423–434

Whittle P (1954) On stationary processes in the plane. Biometrika 41(4):434–449

Zhang G, Taniguchi M (1995) Nonparametric approach for discriminant analysis in time series. Journaltitle Nonparametric Statist 5(1):91–101

Index

A
ARMA process, 8, 30, 106, 109–111, 116
AR process, 33, 51, 53–57, 82–85, 93, 106, 109, 110, 112, 119
Asymptotically normal, 44, 60, 80
Autocovariance function, 3

C
Characteristic function, 2
Check function, 59–61, 66
Confidence region, 87, 89, 90, 105
    confidence interval, 107, 108
Consistency, 47, 60, 63, 66, 84, 93
Contamination, 18, 52
Contrast function, 29–32
Cumulant, 42, 47–49, 52, 53, 62, 81, 90, 97, 98
CUSUM, 119, 120

D
Dirac delta function, 74
Disparity, 30–39, 42, 44–47, 49–54, 56
    location disparity, 31
    scale disparity, 31

E
Efficiency, 29, 47, 49, 52–54
Empirical likelihood, 87–89, 91–93, 103, 106–108
Extrapolation problem, 1, 9–12, 18–22

F
Fisher information, 45, 49, 50, 53

Frequency domain, 1, 5, 11, 54, 59, 60, 73, 80, 81, 87, 88

G
Gaussian process, 2, 29, 30, 32, 45, 49, 53, 82–84
Generalized empirical likelihood, 109
    GEL, 109, 111–114, 117–121

H
Hardy space, 12, 13
Heaviside step function, 74
Herglotz's theorem, 4, 61
Hill's estimator, 105
Hypothesis, 59, 81, 87, 90, 94, 110–112
    change point hypothesis, 116, 118, 119

I
Independent and identically distributed, 9, 11, 36, 46, 88, 91, 94
Infinite variance, 9, 29, 39, 46, 47, 53, 55, 93
Interpolation problem, 1, 9–11, 16–20, 22, 24–26, 32–34, 36, 49

J
Joint distribution, 1–3, 29, 67, 70, 73
Jump, 75

K
Kolmogorov's formula, 6, 15, 35

Printed in the United States
By Bookmasters